WITHDRAWN

GIS for

Everyone

Exploring
Your Neighborhood and Your World with a
Geographic Information System

David E. Davis

ESRI PRESS

Published by
Environmental Systems Research Institute, Inc.
380 New York Street
Redlands, California 92373-8100

Copyright © 1999 Environmental Systems Research Institute, Inc.
All rights reserved. Printed in the United States of America.

The information contained in this document is the exclusive
property of Environmental Systems Research Institute, Inc.
This work is protected under United States copyright law and
the copyright laws of the given countries of origin and applicable
international laws, treaties, and/or conventions. No part of this
work may be reproduced or transmitted in any form or by any
means, electronic or mechanical, including photocopying or
recording, or by any information storage or retrieval system,
except as expressly permitted in writing by Environmental
Systems Research Institute, Inc. All requests should be sent to
Attention: Contracts Manager, Environmental Systems Research
Institute, Inc., 380 New York Street, Redlands, California
92373-8100 USA.

The information contained in this document is subject to change
without notice.

Restricted/Limited Rights Legend
Any software, documentation, and/or data delivered hereunder is subject to the terms of the License Agreement.
In no event shall the U.S. Government acquire greater than
RESTRICTED/LIMITED RIGHTS. At a minimum, use, dupli-
cation, or disclosure by the U.S. Government is subject to
restrictions as set forth in FAR §52.227-14 Alternates I, II,
and III (JUN 1987); FAR §52.227-19 (JUN 1987) and/or
FAR §12.211/12.212 (Commercial Technical Data/Computer
Software); and DFARS §252.227-7015 (NOV 1995) (Technical Data) and/or DFARS §227.7202 (Computer Software),
as applicable. Contractor/Manufacturer is Environmental
Systems Research Institute, Inc., 380 New York Street,
Redlands, California 92373-8100 USA.

ESRI and ArcView are trademarks of Environmental Systems
Research Institute, Inc., registered in the United States and
certain other countries; registration is pending in the European
Community. ArcInfo, ArcExplorer, ArcWorld, ArcAtlas, AML, the
ESRI globe logo, and the ArcExplorer logo are trademarks, and
ArcData and www.esri.com are service marks of Environmental
Systems Research Institute, Inc.

The Microsoft Internet Explorer logo is a trademark of Microsoft
Corporation. The names of other companies and products
mentioned herein are trademarks or registered trademarks of
their respective trademark owners.

Environmental Systems Research Institute, Inc.
GIS for Everyone
ISBN 1-879102-49-8

C O N T E N T S

PREFACE

GIS for Everyone IS NOT A BOOK, even though that's precisely what you seem to be holding in your hand. Rather, it's an interactive introduction to the world of geographic information systems, known familiarly as GIS, that combines printed material, a multimedia CD, and the Internet.

Combined, this package contains a great deal more information than could ever be stuffed into a book this size. Moreover, instead of telling you what to do, as a more conventional "instruction manual" might, *GIS for Everyone* invites you to explore this new world on your own and at your own pace, using all three elements.

We think this will lead you to a better understanding—of what GIS is all about and how you can use it in your life—than a book could provide by itself.

Some of the things you'll be able to do using *GIS for Everyone*:

- Install fully operational GIS software, called ArcExplorer, on your PC. The ArcExplorer™ program isn't a demo that will stop working after a short time. It's yours to install and use for as long as you want.
- Download valuable geographic and demographic data for the ZIP Code where you live.
- View a multimedia gallery of dazzling, informative maps from around the world that were created with GIS software—giving you just a small sample of the power and versatility of this technology.
- Work with more than 500 megabytes of real-world geographic and demographic data on the CD.
- Find links to other GIS sites and data, as well as to software updates and utilities, on the special *GIS for Everyone* Web site. Long after you've finished the book, you'll find this Web site continuing to evolve with new information, updates, and links.

There are a couple of things you'll need to experience *GIS for Everyone* fully: access to a PC and to the Internet, as well as knowledge of some basic desktop computing operations, such as copying and pasting files and folders.

We wish you good sailing as you explore this new GIS world.

A C K N O W L E D G M E N T S

THIS BOOK could not have been written without the cooperation of the organizations that shared their data for the explorations, Map Gallery, and sample data on the CD.

The San Diego Association of Governments (SANDAG) provided most of the data for exploration 1. Other San Diego data layers can be downloaded from the SANDAG Web site at **www.sandag.cog.ca.us**.

Geographic Data Technology, Inc. (GDT), provided the data for Washington, D.C., Rio de Janeiro, and Austin. GDT data is featured in many of the Map Gallery images. You can download several GDT layers with the special access code inside the back cover of this book and visit the GDT Web site at **www.geographic.com**.

Horizons Technology, Inc. (HTI), provided its Sure!MAPS® RASTER topographic map data for San Diego, samples on the CD, and throughout the Map Gallery. You can visit HTI's Web page at **www.horizons.com**.

The Texas Natural Resources Information System (TNRIS), a division of the Texas Water Development Board, provided the aerial image of Austin, Texas, used in exploration 10. You can access and download digital geographic data for Texas at the TNRIS Web site at **www.tnris.state.tx.us**.

The Institute of Municipal Informatics of Capital Prague provided all the background data for exploration 8 such as streets, parcels, and water. ARCDATA PRAGUE provided the restaurant and attractions themes used in the same exploration. Special thanks to Jan Vodnansky and Petr Seidl. You can visit their Web site at **www.arcdata.cz**.

Several other organizations provided data specifically for use in the Map Gallery or sample data. Thanks to Geosystems Romania, The MapFactory, Tele Atlas B.V., VISTA Information Solutions, Inc., AND Mapping B.V., the National Oceanic and Atmospheric Administration (NOAA) Coastal Services Center, and WorldSat International.

Special thanks to Jack Dangermond, president of ESRI, for his societal GIS vision and desire to bring GIS to everyone.

Next, I thank Bill Miller and Judy Boyd for allowing me to write this book and providing the best support imaginable. Christian Harder, manager of ESRI Press, gave me his enthusiasm and trust and led me through the book-publishing process.

And to all my colleagues in the Educational Products department, thank you for generously sharing your time and professional experience throughout the project.

Thanks to R.W. Greene and Michael Karman for carefully editing each page of the manuscript. Your contributions to the book are invaluable.

Michael Hyatt designed the book and did all page production, copyediting, and proofreading. Tammy Johnson designed and produced the cover. Barbara Shaeffer reviewed the text from a legal perspective.

My sincere thanks to Deane Kensok for his many ideas and for the hard work he put into this project, including obtaining data and permissions. He and Anne Wilson developed the companion Web site.

Clem Henricksen and Rick Schneblin created the Map Gallery and ArcExplorer video content on the companion CD–ROM.

Judy Boyd, Mike Tait, Deane Kensok, Charlie Fitzpatrick, and Michael Phoenix read drafts of the book and provided excellent comments.

Robin, Jerry, and Mark Poupard tested the many interactive explorations in the book.

Introduction

IF YOU PICKED UP THIS BOOK to browse, it's possible you did so because your eye was attracted to the maps on the cover, or you were curious about the phrase "geographic information system."

Or maybe not.

But it's entirely possible that you're one of those people who dreamed as a kid about visiting far-off lands, and that you have a desk drawer somewhere stuffed with *National Geographic* maps dating back to 1968—not to mention a road atlas so tattered and beat up only you can read it.

It's possible as well that you've bookmarked a half-dozen sites on your Internet browser that let you do things like chart hurricanes, or that bore your kids with school report facts about the surface area and exports of France.

Or maybe you're one of those people who just like knowing stuff about the world around you, even if you haven't looked at a map in years.

No matter which group you fall into, you're in luck, and not just because you put your hands on this book.

You're in luck because you live at a time when the quality and quantity of information about the world around you—geographic information—is expanding at an astonishing rate.

It's possible now to make maps in minutes, from information freely available on the Internet—maps so packed with useful information accessible with only a few mouse clicks, that they're no longer just maps.

Rather, they're smart maps, made possible by geographic information systems (GIS), and everyone, including people who never look at a map, will find they can make life much easier and more interesting.

GIS has been around for years, but the digital revolution has created, and continues to create, many new applications for the technology. GIS is now used by industries and governments worldwide in ways unimaginable only a few years ago.

GIS technology is helping people make better decisions in a host of areas, such as agricultural and natural resources management, and environmental control. It's helping businesses streamline customer service operations, coordinate enterprisewide problem-solving, and revolutionize logistical planning. It's saved millions of dollars through increased productivity and efficiencies.

A GIS is a kind of supermap, computer software that links geographic information (where things are) with descriptive information (what things are like). Unlike a flat paper map, where "what you see is what you get," a GIS can have many layers of information underneath its surface.

Moreover, that descriptive information is virtually unlimited in both depth and breadth.

If you look at a road on a paper map, about all you see is a name and maybe a highway number. If you click on the same road on a GIS map, you might find not only its name, but also how many lanes it has, when it was built, what the road surface is made of, when it was last painted, and whether you can see that spot on the road from a mountain 20 miles away.

With the right software, you can even create an animated scene in which you're flying down the road as if in a helicopter.

And everyone can use a GIS. It's not just technology for industries and governments.

You can use a GIS at home—to show the most scenic route to a vacation spot, to draw maps to a garage sale or for school reports, or to chart the housing prices and SAT scores in an area where you're thinking of buying a new home.

You can use a GIS at work—to chart where your best customers are likely to live, where the cheapest office space is, or just to find out how many Italian restaurants there are within ten minutes of the office.

Most of this kind of information is a lot easier to get than you might think. Local and state governments, as well as the federal government, provide the bulk of it, and those governments put it on the Internet. What they don't put on the Internet is usually available down at city hall or the county courthouse or the local library.

And most of the time, since it's public information, it's free.

A tool that allows you to explore all this digital information, a software program called ArcExplorer, comes free on the CD included with this book. It does a lot and is easy to learn. Once you've mastered it, we can show you how to get even more powerful software.

In this book, we'll introduce you to some of the basic ideas behind creating digital maps, and then with the help of the CD, you'll put these ideas to work by following simple step-by-step instructions. You'll use real information from real cities around the world. By the end of the book, you'll be able to create stunning digital maps for any purpose under the sun, using data from sources all over the Internet. You'll be hooked on GIS.

And by the way, for those of you who just like knowing stuff about the world around you, the surface area of France is 547,000 square kilometers, and its principal exports are machinery, transportation equipment, and of course, wine.

CHAPTER 1

GIS for everyone

In this chapter, you'll get a sense of the enormous variety of digital maps that people are making, and making available for free, on the Internet right now. It's that variety that makes GIS as powerful as it is; you're really only limited by your own imagination.

To see the show, you can either flip through the Map Gallery on the next few pages to see what a GIS can do, or put the CD into your PC, following the instructions below.

ACCESS THE MAP GALLERY ON CD

1 Insert the *GIS for Everyone* CD in your CD–ROM drive.

2 Choose Run from the Start menu.

3 In the Command Line box, type the letter of your CD–ROM drive, a colon, a backslash, and gallery.exe (for example, **e:\gallery.exe**).

Map Gallery

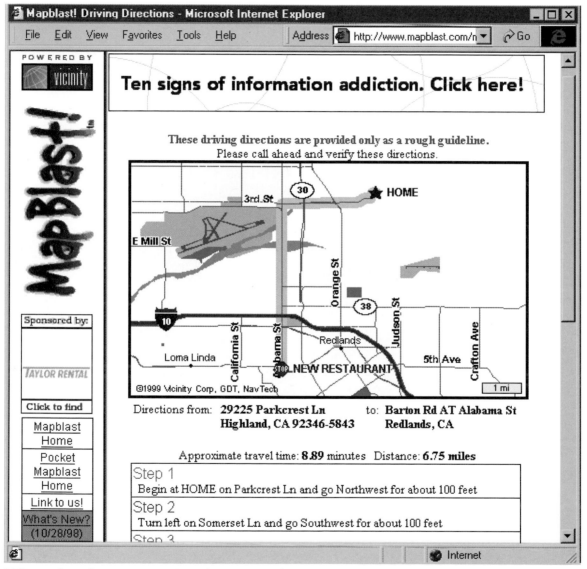

Millions of people access geographic information on the Internet every day. They go to sites such as MapBlast to create custom maps of almost any place in the United States. Here you see a map of driving directions.

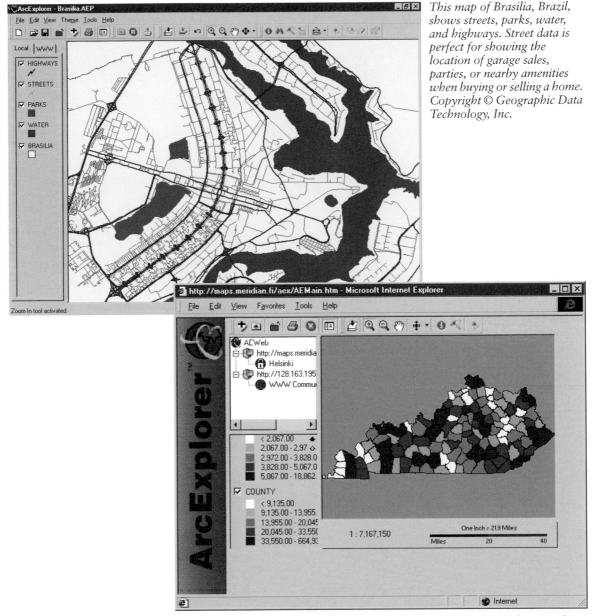

This map of Brasilia, Brazil, shows streets, parks, water, and highways. Street data is perfect for showing the location of garage sales, parties, or nearby amenities when buying or selling a home. Copyright © Geographic Data Technology, Inc.

Some Web sites allow you to access maps using a special version of ArcExplorer that runs in your Web browser. Here you see the WWW Community GIS site of the Social, Natural and Agricultural Resources Information Laboratory at the University of Kentucky (snril.ca.uky.edu). This map of Kentucky shows population by county.

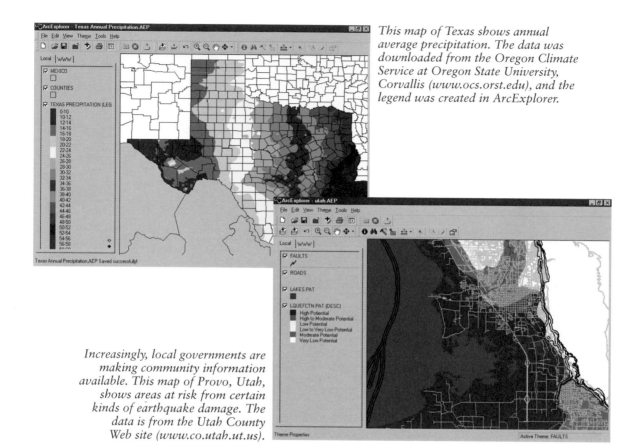

This map of Texas shows annual average precipitation. The data was downloaded from the Oregon Climate Service at Oregon State University, Corvallis (www.ocs.orst.edu), and the legend was created in ArcExplorer.

Increasingly, local governments are making community information available. This map of Provo, Utah, shows areas at risk from certain kinds of earthquake damage. The data is from the Utah County Web site (www.co.utah.ut.us).

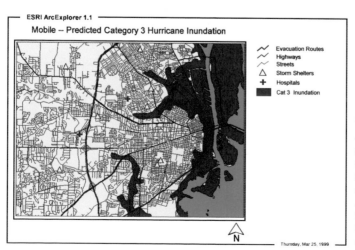

This map of Mobile, Alabama, shows the area forecasters say could be in danger from the storm surge of a category 3 hurricane. Data is from the NOAA Coastal Services Center, Charleston, South Carolina.

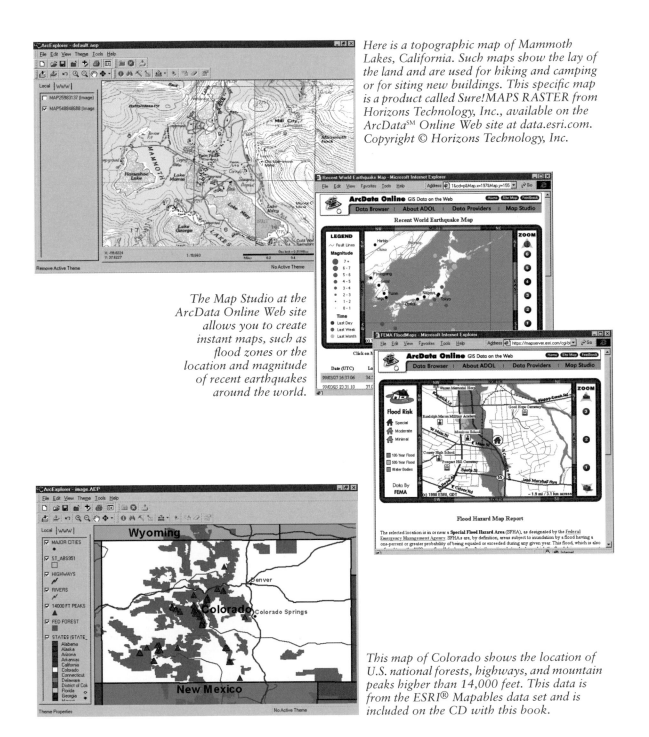

Here is a topographic map of Mammoth Lakes, California. Such maps show the lay of the land and are used for hiking and camping or for siting new buildings. This specific map is a product called Sure!MAPS RASTER from Horizons Technology, Inc., available on the ArcData℠ Online Web site at data.esri.com. Copyright © Horizons Technology, Inc.

The Map Studio at the ArcData Online Web site allows you to create instant maps, such as flood zones or the location and magnitude of recent earthquakes around the world.

This map of Colorado shows the location of U.S. national forests, highways, and mountain peaks higher than 14,000 feet. This data is from the ESRI® Mapables data set and is included on the CD with this book.

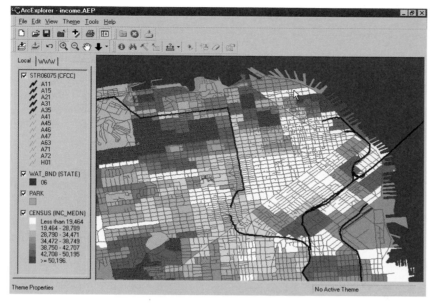

This map of San Francisco shows the median income of households for groups of city blocks. A business might target certain areas for advertising. This data is available free of charge at ArcData Online, data.esri.com.

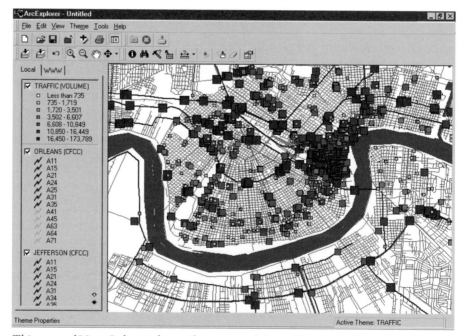

This map of New Orleans shows the average daily traffic volume for selected street locations. This data, from Geographic Data Technology, Inc., is available at ArcData Online. Copyright © Geographic Data Technology, Inc.

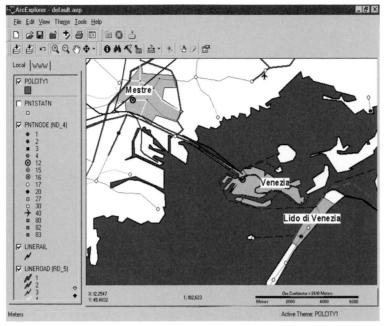

This map of Venice, Italy, shows transportation features such as highways, railroads, airports, and ferry crossings. This data is from AND Mapping of Rotterdam, The Netherlands, and is available for downloading at ArcData Online. Copyright © AND Mapping.

This map of Los Angeles shows streets and buildings. The data set is from The MapFactory and is included on this book's CD. Copyright © The MapFactory.

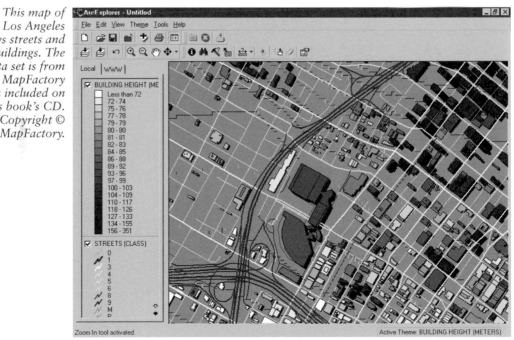

Understanding digital maps

TO USE A PAPER MAP, all you have to do is unfold it. You have to take over the entire dining room table to do this, of course, but once you do, spread out before you is a magnificent representation of cities and roads, mountains and rivers, railroads and state lines. The cities are represented by little dots or circles, the roads by black lines, the mountain peaks by tiny triangles, and the lakes by small blue areas similar in shape to the real lakes.

A digital map is not much more difficult to use than a paper map, and it takes up a lot less space. As on the paper map, there are dots, or points, that represent features on the map such as cities; there are lines that represent features such as roads, and small areas that represent features such as lakes. Everything is neatly labeled. The colors are just as bright.

But a digital map takes up less space because all that information—about where the point is, and how long the road is, and how many square miles the lake occupies—is stored in digital format, as a lot of ones and zeros.

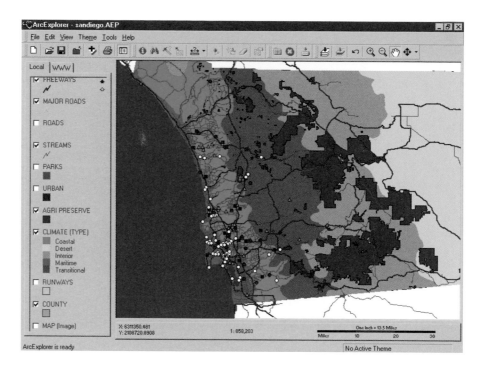

It's helpful to think of this digital geographic data as layers of information underneath the computer screen on which you look at the map. These layers are called *themes,* and each theme represents a particular feature of the area you're looking at. One theme could be made up of all the roads in an area. Another theme could represent all the lakes in the same area. Yet another could represent all the cities.

These themes can be laid on top of one another, creating a stack of information about the same geographic area. They can be turned off and on, as if you were peeling a layer off the stack, or placing it back on. You control the amount of information about an area that you want to see. If you turn off all the layers, you'll get a blank screen. If you turn on the cities theme, the roads theme, the lakes theme, the mountains theme, the rivers theme, the railroads theme, and the state boundaries theme, you have a map on your computer screen that pretty much matches the paper map you started with that was spread out all over the dining room table.

ArcExplorer, the software used to create the maps in chapter 1, is included on the companion CD, and it's high time you installed it. Then we'll proceed to explore some themes in the San Diego area.

INSTALL ARCEXPLORER

To run ArcExplorer, you must have Microsoft® Windows 95®, Windows 98®, or Windows NT® version 4.0 installed on your system. If you're running Windows NT, you must have Service Pak 3 installed.

To install ArcExplorer,

1 Insert the *GIS for Everyone* CD in your CD–ROM drive.

2 Choose Run from the Start menu.

3 In the Command Line box, type the letter of your CD–ROM drive, a colon, a backslash, and the file name aeclient.exe (for example, **e:\aeclient.exe**).

4 When asked which components to install, choose all of them:

◆ Application Files: The core of ArcExplorer software.

◆ Help and Tutorial Files: Online help system and an ArcExplorer manual in PDF format (aemanual.pdf).

◆ Web Integration Tools: The World Wide Web (WWW) functionality of ArcExplorer. These tools allow you to view and download data from the Web using ArcExplorer.

The setup program will automatically install ArcExplorer on your PC and place a shortcut to start the program on your desktop.

You'll find an ArcExplorer toolbar reference page at the back of this book.

Exploration 1   Look at San Diego

In this first exploration you see a paper map, from the United States Geological Survey, that's been scanned into the computer. On it, you can find the San Diego Zoo, the airport, and Sea World, just as you would with a paper map. You'll look at the same area on a digital map to get your first taste of how these things are represented as layers of digital geographic data.

1 Start ArcExplorer either by double-clicking the ArcExplorer shortcut on your desktop or by selecting Programs:ESRI:ArcExplorer from the Start menu. You'll see the ArcExplorer opening banner, then the ArcExplorer window.

2 Click the Open Project button. In the dialog box that displays, navigate to the *explore\sandiego* directory on the CD.

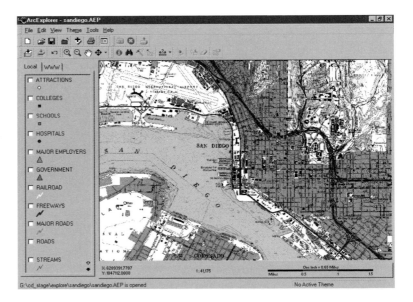

3 Select the project file called *sandiego.AEP* and click Open. (The AEP extension stands for ArcExplorer Project.) Project files store all the work you do with ArcExplorer. They contain the paths to the data and information about how it's displayed.

In the map view is a scanned paper map of San Diego. It shows such features as elevation, highways, streets, parks, buildings, airports, and bodies of water. Scanned maps are excellent visual backdrops for other geographic data. Although you can learn a lot from a paper map, the amount of information you can get from it is limited to what's shown. And you can't exclude things you may not need to see.

In ArcExplorer, each layer of digital data—the theme—is listed to the left of the map view in what is called the legend. The map view of San Diego contains a theme for streets, one for highways, another for parks, as well as additional themes for major attractions, hospitals, and other features of life in San Diego.

4 Use the down arrow at the bottom of the legend to scroll down. In the legend, turn off the "paper" map by unchecking the box next to the MAP theme. This turns off the view of the scanned paper map. Turn on the RAILROAD, FREEWAYS, MAJOR ROADS, ROADS, PARKS, RUNWAYS, and COUNTY themes by checking the boxes.

This looks familiar. But in this digital map view of San Diego, you can look at different features and combinations of features in the area by turning themes on and off. With the paper map, what you see is what you get. Moreover, in the digital map, you can perform particular operations on a theme by making it "active." You do this by clicking on its name in the legend. When it's made active, the theme will appear to be raised above the surface of the legend, and you'll see its name at the bottom right of the ArcExplorer screen.

5 Turn on the ATTRACTIONS, COLLEGES, SCHOOLS, HOSPITALS, MAJOR EMPLOYERS, and GOVERNMENT themes by checking their boxes in the legend. Suddenly, a bunch of squares, circles, and triangles appear, each representing a different theme.

6 Click on the name ATTRACTIONS in the legend to make the theme active. This makes additional operations with the theme possible.

7 Move your mouse pointer over some of the yellow circles. Because the theme is active, the name of the attraction will appear above the circle when you pass the pointer over it. See if you can find the San Diego Zoo, which is in a large park near the upper right of the map view. In the graphic below, we've found San Diego's international airport, also known as Lindbergh Field.

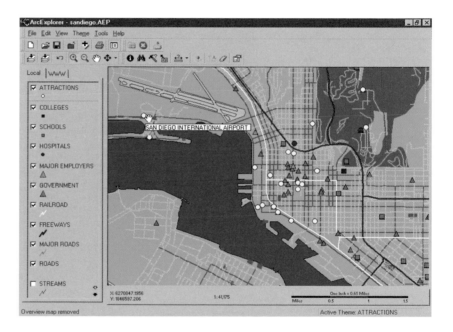

Try exploring some other themes. Click a theme name to make it active. You have the choice of HOSPITALS (red circles), GOVERNMENT (orange triangles), COLLEGES (blue squares), SCHOOLS (orange squares), or MAJOR EMPLOYERS (light blue triangles). Move your mouse pointer over some features to see their names.

Now that you've identified some features, it's time to move around the map. First, you'll explore San Diego with the Direction button.

8 Choose a direction using the down arrow at the right of the Direction button. An arrow appears on the button to indicate the direction you selected. Click the Direction button to move in that direction. Try out a couple of other directions as well. Don't forget to drag your mouse pointer over things you encounter along the way. (And don't forget to make the theme active.)

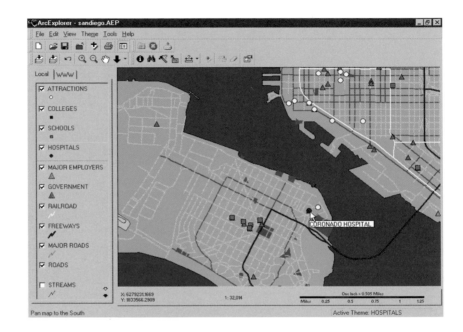

Notice that your map view changes in easy, controlled steps. Here you see that we panned south.

Try the Pan button. With it, you grab your display and drag it in any direction.

9 Click the Pan button. Move your mouse pointer into the map view, hold down the mouse button, and drag. When you get to where you want to be, release the button.

We're not sure where you'll end up in this exploration. Go ahead and move in any direction you want. In the graphic above, we panned to the northwest and ended up in the vicinity of Sea World.

Some themes represent features located all over the map; some represent features located in one area. In either case, you can zoom to the area covered by a particular theme. To see the whole map, the area covered by all the themes, you can click the Zoom to Full Extent button. Or you can zoom to the active theme with the Zoom to Active Theme button. And of course you can zoom in and out. Don't worry if you get lost. That Zoom to Full Extent button will bring you back to a view of the entire map.

10 Click the Zoom In button. Click once somewhere in the map view to zoom in on that area. Click again to zoom in even more. You can also drag a box over an area to zoom in on it, as shown here.

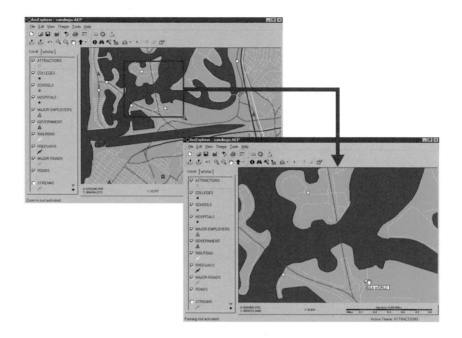

11 Click the Zoom Out button. Click once somewhere in the map view to zoom out from that location. Click the Zoom Out button once more.

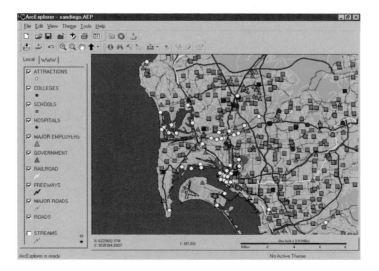

It's neither practical nor possible to include all the information about a place on a single map. So navigating in a digital map goes beyond zooming and panning with what you see. You have access to several different maps of the same place, and you can get to them as you navigate.

12 Turn off all themes except COUNTY.

13 Zoom to the extent of the COUNTY theme. (Make that theme active by clicking on its name, then use the Zoom to Active Theme button.)

Now you see all of San Diego County. The city of San Diego is a coastal town, but the much larger county is mostly farmland, with the glamorous avocado a specialty.

14 Turn on the AGRI PRESERVE theme.

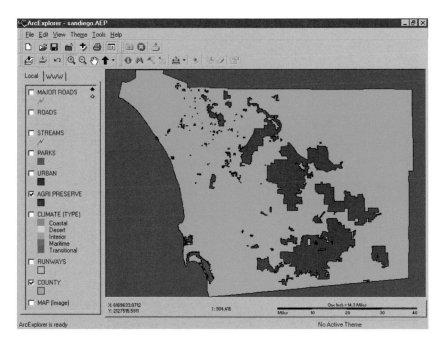

What about the climate for all that agriculture?

15 Turn on the CLIMATE theme. You can see how many climate zones the county has, from the cool coastal climates in the west, to the hot, dry desert climates farther east.

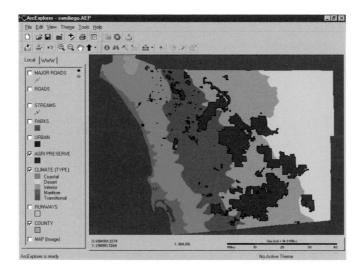

Perhaps now you're wondering what other effects the climate has.

16 Turn off the AGRI PRESERVE theme. Turn on the URBAN theme. Just like that, you've changed the subject of your map. And you can see that most people would rather live along the coast than in the desert. Housing prices reflect that, although that's not a theme here.

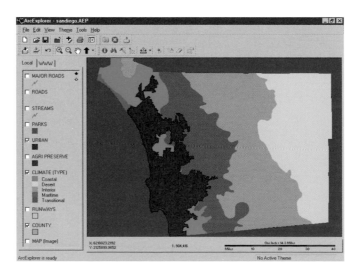

17 Click the Close Project button. Choose No when asked if you want to save any changes.

Now you've had a chance to see how a digital map—a GIS—is made up of themes, and how you can navigate through those themes with your mouse, or by turning themes on or off, or making them active. These capabilities make a GIS a powerful tool. There's one thing a GIS can't do, however: figure out how to fold up that paper map.

CHAPTER 3

Finding answers with digital maps

UPROOTING YOUR FAMILY and moving to a new city might be more fun than a tax audit, but not by a wide margin. Nevertheless, in an era when economies are no longer local or regional, but global, it's an event increasingly common in the lives of many.

Think of the questions that flood your mind when you first find out you're moving someplace new. What are the neighborhoods like? Can we afford a house in a nice neighborhood? Are housing prices reasonable? How far will I have to commute? Are the schools any good? Is there some place nearby where we can go hiking? Can you get a decent bagel there?

Pulling out a paper map of the new town might give you an answer to your commute question, but that's about it. You'd fold up the map and put it back in the glove box to gather more road dust.

With a GIS map of the new town opened up on your computer monitor, you could get answers to all those questions, except perhaps the one about the bagels. Answering questions is one of the things a GIS does best.

In the last chapter, you learned that you can think of digital geographic data, the ones and zeros, as being in a series of layers behind a map view. All the roads are in one layer, all the highways in another layer, all the schools in another, and so on.

Those layers are actually made up of two kinds of information about each geographic feature—its location on the planet and its description.

The files that store location are the ones that allow you to see features on your screen such as streets, buildings, and parks. The description of a feature is known as its attributes. Attributes are stored in tables in rows and columns. Attributes can be as numerous and complex as you want them to be. For example, the attributes in a theme of streets might include a street's name, its overall length, the speed limit along it, the route number, the road material, when it was built and by which contractor, and the number of lanes. This information would be replicated in the table for each street in the theme.

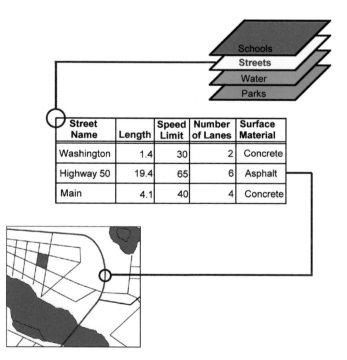

Street Name	Length	Speed Limit	Number of Lanes	Surface Material
Washington	1.4	30	2	Concrete
Highway 50	19.4	65	6	Asphalt
Main	4.1	40	4	Concrete

In this chapter we'll look at how you can use the information in these tables to answer all kinds of questions.

Exploration 2   What is that?

Back in San Diego, during exploration 1, you swept your mouse pointer over various features and saw their names pop up on your monitor. The tool that does this is called the MapTips tool. You can set the MapTips tool to read any of the attributes of a feature, because a feature's name is only one of many possible attributes.

ArcExplorer

1 Start ArcExplorer, if necessary.

2 Click the Open Project button. In the dialog box that displays, navigate to the *explore\dc* directory on the CD.

3 Select the project file called *dc.AEP* and click Open. When the project opens, you see a map view of downtown Washington, D.C., with themes of streets, highways, parks, landmarks, water, and institutions.

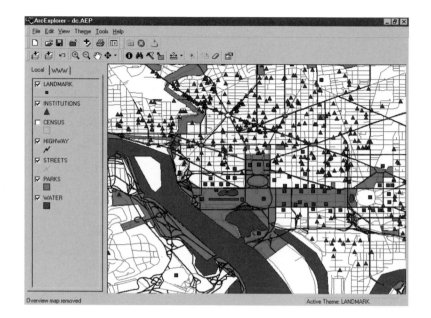

4 Make the LANDMARKS theme active.

5 Click on the MapTips tool. The MapTip Field Selection dialog box displays. This allows you to choose which attribute you want to pop up when you pass the mouse pointer over a particular element.

6 Choose the *NAME* field and then click OK in the dialog box. The name of a particular landmark will appear when you pass your mouse pointer over it. It doesn't matter if the pointer is also set to do something else, like zooming in or panning. It will do both tasks.

7 Drag the mouse pointer over other landmarks and read their names.

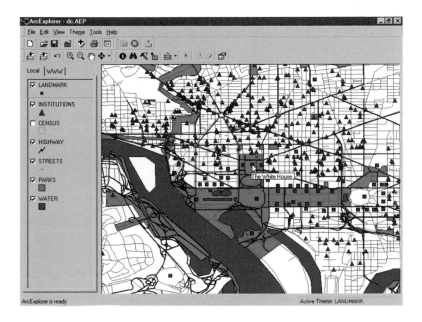

Changing the field you want to pop up is a simple matter.

8 Click on the MapTips tool. Choose the *ADDRESS* field and then click OK in the dialog box.

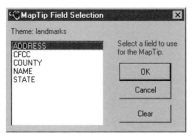

9 Drag the mouse pointer over the landmarks. Now, the address or some kind of location information about each landmark appears.

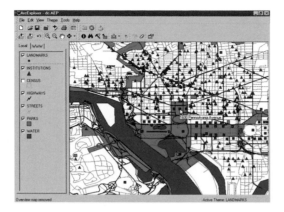

Another method of discovering information about a theme is to use the Identify tool.

10 Turn off the STREETS, HIGHWAYS, INSTITUTIONS, and LANDMARKS themes. Turn on the CENSUS theme and make it active.

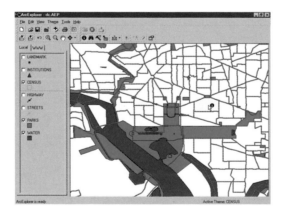

The CENSUS theme contains information about people—where they live, where they were born, how much they earn, what education level they attained, how old they are, how much rent they pay, their ethnicity. Census data is in fact one of the richest sources of information about the United States, and is easily accessible.

Census data is commonly divided into block groups, which are simply a collection of city blocks. You can use the Identify button to learn something about the people who live in various parts of Washington, D.C., from the census data of block groups.

11 Click the Identify button. Now move your mouse pointer into the map view and click on some areas of the city at random.

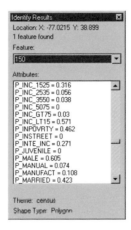

The first time you click on an area, the Identify Results dialog box will display. It lists all the attributes associated with the block group you clicked.

It's difficult to tell anything about the people of Washington, D.C., immediately. In fact, the entries in this box look like gibberish. They are, however, simply abbreviations that census takers have invented for various categories. For example, the first entry means "Able to speak English." If you scroll down the list, you'll probably be able to figure out most of the others. For a full list of abbreviations translated into English, consult appendix B.

12 Close the Identify Results dialog box by clicking the X in its upper right corner.

13 Turn off the CENSUS theme when you finish exploring its attributes. Turn the STREETS, HIGHWAYS, INSTITUTIONS, and LANDMARKS themes back on.

Exploration 3 Where is it?

MapTips and Identify give you information about places you point to on the map. But sometimes you'll already have information about a place, and you want to find it on the map. To do this, you'll use the Find tool.

1 Click the Find button. The Find Features dialog box displays.

Find Features (Text searches only)

1. Enter the text you want to find (searches are case-sensitive)

2. Select a search type
   ● Any Part of Field   ○ Whole Field   ○ Start of Field

3. Choose which themes to search

   landmark
   Institutions
   census

   Find

4. Pick a feature   0 matches found

Theme	Feature	Value

5. Select the operation to perform

   Highlight   Pan To   Zoom To

---

2 Type **Smithsonian** in the text box. Note that the Find tool is case-sensitive, so be sure to enter the text exactly as shown.

3 Since we didn't enter the full name of this famous museum— Smithsonian Institution—choose *Any Part of Field* in section 2 of the dialog box.

4 Choose *LANDMARKS* as the theme to search in section 3.

5 Click the Find button. ArcExplorer proceeds to search the features in the LANDMARKS theme and returns a list of matches.

6   Click on the match for the Smithsonian Institution to highlight it. You'll see that the Highlight, Pan To, and Zoom To buttons are no longer grayed out.

7   Click the Highlight button. The point representing the Smithsonian on the map will flash briefly.

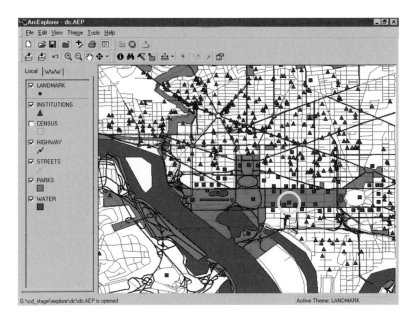

8 Click the Zoom To button in the Find Features dialog box. You're zoomed in to the Smithsonian Institution, its location flashes, and it's now in the center of the map view.

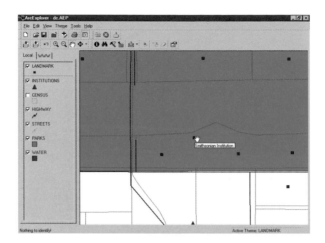

9 Click the Zoom to Full Extent button. You see the entire map—that is, the area covered by all the themes in the legend.

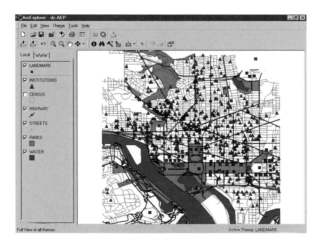

Like all of the tools you've used so far, the Find tool works with any theme.

10 Type **CONSTITUTION GARDENS** as the text you want to find in the Find dialog box. (Be sure to enter the text in all capital letters.)

11 Choose *Any Part of Field* as the search type.

12 This time, choose the *PARKS* theme as the theme to search.

13 Click Find.

14 Click on the match for Constitution Gardens.

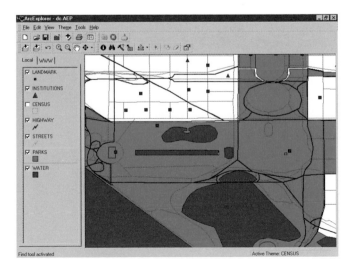

15 Use the Highlight button to make the park's location flash. Click the Zoom To button in the Find Features dialog box to zoom in on the park.

16 Close the Find Features dialog box.

Exploration 4   How far is it?

Now that you've gotten an idea of what's in the data, you're ready to perform some real geographic analysis, like using that data to measure distances.

1 Click the Zoom To Full Extent button.

2 Zoom in on the area shown in the red box below with the Zoom In tool.

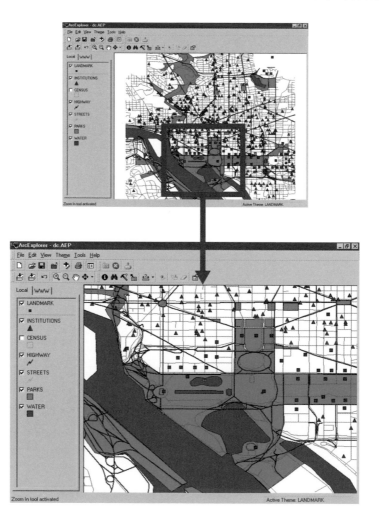

Finding distances is a two-step process. First, you must tell the GIS what kind of measurement units your map data is stored in. Second, you need to tell it which kind of measurement units you want it to use to tell you the distance between one place and another.

The measurement units that map data is stored in are known as map units. Because there are so many different ways to collect and store data, it would be impossible for ArcExplorer to determine which map units were used in a project. So to get an accurate measurement, you must tell ArcExplorer which map units your data is stored in. If you don't know the map units for your GIS data, check any accompanying documentation that came with the data, or ask the person who gave it to you. (If you can't find this information, you can zoom in on a familiar area and try all the possible map units until you get one that seems to provide the most accurate results. It would be better, of course, to obtain the map units from another source.)

The ArcExplorer default map units are decimal degrees, and all of the Washington, D.C., themes use this default. Decimal degrees are degrees of latitude and longitude expressed as decimals rather than as degrees, minutes, and seconds. (For example, a point located at longitude 73 degrees, 59 minutes, and 15 seconds would be expressed as 73.9875 decimal degrees.) To use the Washington, D.C., data, you need only make sure that ArcExplorer is indeed set to decimal degrees.

As for the units that you want ArcExplorer to use in telling you how far something is—the distance units—you have your choice of feet, miles, meters, and kilometers.

3 From the View menu, select Scale Bar Properties, then Map Units. Make sure that *Decimal Degrees* is checked. (If you use data that has different units, this is where you would tell ArcExplorer.)

Map Units ▶	✓ Decimal Degrees
Scale Units ▶	Feet
Screen Units ▶	Meters

4 Use the Find tool to locate the Lincoln Memorial, Washington Monument, and Jefferson Memorial if you don't know where they are. Dismiss the Find box after you've found them.

5 Click on the Measure tool down arrow, then choose *Miles* from the list. These are your distance units.

Feet
Miles
Meters
Kilometers

You can either measure distance between two points in a straight line or you can click several points along a route (at each turn) to get the total distance from start to finish.

First you'll measure the distance between the Lincoln Memorial to the Washington Monument.

6 Click on the Measure tool (the pointer changes to a crosshair). Since the direct path between the two landmarks is through water, you'll need to measure the path in several segments. Click and hold the mouse as you drag a line segment from the center of the Lincoln Memorial toward the Washington Monument, avoiding the water. Release the mouse button at the end of your first segment.

The segment and total length you measured are displayed in the status panel at the top left of the map view.

Click and hold your mouse button again to complete your route to the Washington Monument. Release the mouse button at the end of each additional segment. Use as many segments as you need to measure the route.

The distance is about .8 miles from the Lincoln Memorial to the Washington Monument.

7 Click and hold your mouse again as you drag more line segments from the center of the Washington Monument to the center of the Jefferson Memorial, avoiding the water.

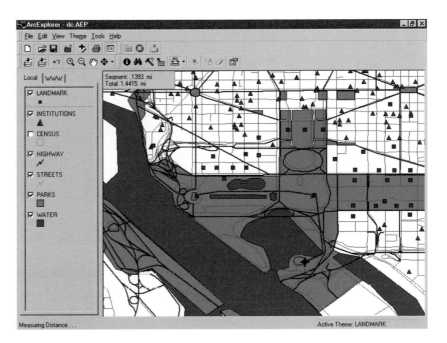

The total distance is about 1.5 miles.

If you want to measure the distance between more things, double-click anywhere in your map view to begin a new measurement. After you double-click, the total length of your previous measurement, in this case the distance from the Lincoln Memorial to the Jefferson Memorial, appears in the lower left corner on the status bar.

Exploration 5 What's it like?

A third way of answering questions about an area is by using the Query Builder. As the name suggests, this tool helps you put together a question that begins "Where is…?" The answer to the question is shown on the map. This tool is especially suited to working with numeric attributes such as those found in census and demographic data. That data is commonly used by businesses to find customers.

Suppose, for example, that you wanted to open a small coffeehouse, specializing in boutique coffee drinks such as espresso and cappuccino. You'd want a location where people needed coffee. You'd also want a location where they could afford your high prices. You'd use the Query Builder to help you.

1 Click the Zoom to Full Extent button.

2 Turn off the STREETS, HIGHWAYS, INSTITUTIONS, and LANDMARKS themes. Turn on the CENSUS theme and make it active.

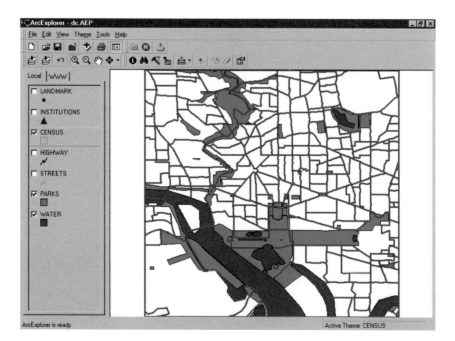

First, you need areas of the city where people who can afford your coffee live; say, people who make more than $30,000 a year. You can find these areas by writing an expression, called a query, that instructs ArcExplorer to find those areas where the query expression is true.

3 Click the Query Builder button. The Query Builder dialog box displays.

At the top of the dialog box you see the name of the active theme, Census. The dialog box contains a list of field names (at the left), that is, all the different kinds of demographic data that were collected for this theme; a set of operators (center) that will do the actual work of narrowing down the data; and a list of sample values (right). When you click on a field name, all the unique values for that field display in the Sample Values list.

To build a query, you click on a field name, click on an operator, then click on a value or type it in. As you build the query, it displays in the query text box in the center of the dialog box. (You can also type your query directly in the query text box, but your typing has to be very accurate, or the Query Builder won't perform the query.)

4 In the Query Builder dialog box, scroll down through the list of field names. If necessary, you can check appendix B for a translation of some of the more obscure abbreviations.

Click on *INCPRCAP.* You see the values for this field display in the Sample Values list.

5 Click the greater-than operator (>). It displays in the query text box.

6 Type in the value **30000** at the end of the expression in the query text box. Your expression should look like this: **INCPRCAP > 30000**

In English, this means, "Find all the census blocks in which per capita annual income is more than $30,000."

7 Choose *INCPRCAP* as the Display Field. This will display the values for this field for all the matching records.

8 Click the Execute button. ArcExplorer searches the attribute table for all the records that match your request. Matches for areas having a per capita income greater than $30,000 are shown in the query results section of the dialog.

9 Click the Highlight Results button. All of the matching areas are high-lighted in your map view in bright yellow. Move the dialog window to the side so you can see your map view.

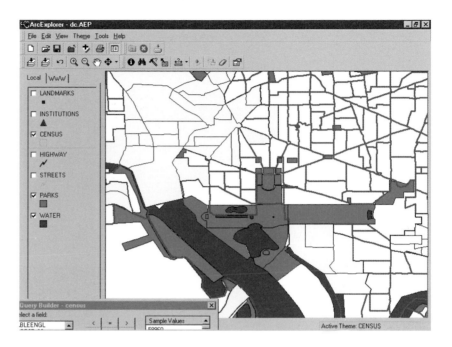

Now you need to find areas where there's a strong need for coffee; say, people with long commutes to work. You need to build an expression that finds the areas where people make more than $30,000 a year and have long commutes.

10 Clear the old expression by clicking the Delete button. Build the following expression in the dialog: **INCPRCAP > 40000 and COM_GT44 > 50**

In English, this means, "Find all the areas where people make more than $30,000 a year and, in the same area, there are more than fifty people who have to drive more than forty-four minutes to get to work."

11 Choose *INCPRCAP* as the Display Field.

12 Click the Execute button. The records that match your query are shown in the query results section.

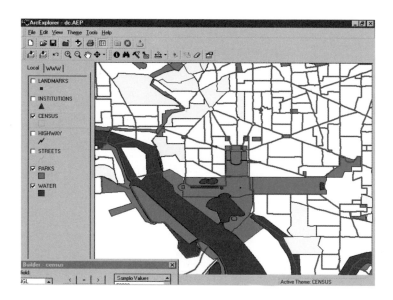

13 Click the Highlight Results button. All of the matching areas are high-lighted in your map view in bright yellow.

Try some other queries. Remember, a translation of the census abbreviations is contained in appendix B.

14 When you're finished, close the Map Query dialog box by clicking on the X in its upper right corner.

Exploration 6   Where is it? (part two)

The last way to find things is to use the Address Matcher tool. This tool allows you to find the location of a particular address, a process known as address matching. It's the computer equivalent of pushing pins into a wall map to show the location of an address. Address matching in ArcExplorer is best used to compare the location of an address with the location of other geographic features, like parks, flood zones, or shopping centers.

Suppose you're headed to Washington, D.C., for business, but have a little free time to see some sights. You have the choice of staying at one of three hotels. You want to see where each is so you can choose the one closest to the major tourist attractions.

First, you need to tell ArcExplorer which theme contains the streets and specify the special attributes that allow address matching to work. Next, you specify an address to look for. ArcExplorer then places a label to show the location of the address on your map.

1   Click the Zoom to Full Extent button.

2   Turn off the CENSUS theme. Turn on the LANDMARKS, STREETS, and HIGHWAYS themes.

3   Make the STREETS theme active.

4 From the Theme menu, choose Address Matcher Properties. The Address Matcher Properties dialog box displays. The required input fields have field names from the STREETS theme filled in. These field names correspond to the default names required by ArcExplorer.

Address Matcher Properties	☒
Input Field:	

∗ Street Name	NAME ▼
∗ Left From Address	L_F_ADD ▼
∗ Left To Address	L_T_ADD ▼
∗ Right From Address	R_F_ADD ▼
∗ Right To Address	R_T_ADD ▼
Left ZIP Code	ZIPL ▼
Right ZIP Code	ZIPR ▼
Street Type	TYPE ▼
Prefix Direction	PREFIX ▼
Suffix Direction	SUFFIX ▼
City	▼
State	▼

OK   Cancel

∗ Required Field

We won't go into an explanation of these fields at this point. If you're interested in learning more about how address matching works, consult the Address Matching topic in ArcExplorer Help.

5 Click OK to make the theme matchable. Once the Address Matcher properties are set, they needn't be set again during your current ArcExplorer session.

Now you can find the location of each hotel address using a street number and name. The three hotel addresses you want to find are:

Holiday Inn Central, 501 Rhode Island Avenue
Capital Hilton, 1001 16th Street
Normandy Inn Hotel, 2118 Wyoming Ave., NW

First you'll find the location of the Holiday Inn and see how close it is to the landmarks.

6 Click the Address Matcher button.

7 In the Address Matching dialog box, choose *Address* under Options and enter **501 Rhode Island Avenue** as the address you want to match.

8 Click the Match button. ArcExplorer finds the address and puts a label on the map. The map view pans and zooms to the location of the Holiday Inn. Move the dialog window to the side so you can see your map view. This hotel is near some landmarks, but perhaps you could be even closer.

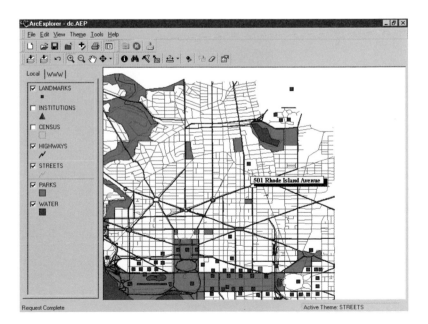

Now find the location of the Capital Hilton.

9 In the Address Matching dialog box, enter **1001 16th Street** as the address you want to match.

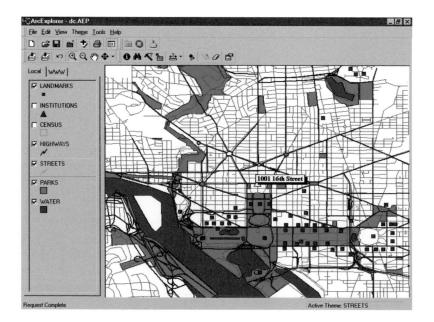

10 Click the Match button. The map view pans and zooms to the location of the Capital Hilton. This hotel offers a very central location in relation to the landmarks. You'll probably stay here, but you'll find the last hotel location just in case.

GIS FOR EVERYONE

Find the location of the Normandy Inn Hotel.

11 In the Address Matching dialog box, enter **2118 Wyoming Ave., NW** as the address you want to match.

12 Click the Match button. The map view pans and zooms to the location of the Normandy Inn Hotel. Like the Holiday Inn, this hotel is too far away. You'll stay at the Capital Hilton.

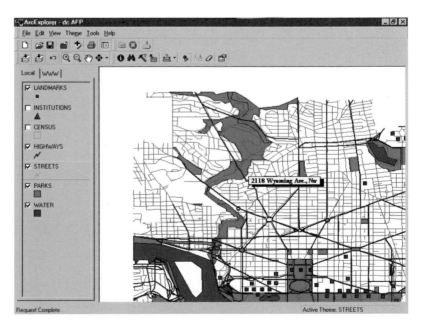

13 Click the Clear Selection button when you're finished viewing the location.

 14 Close the project without saving any changes.

Address matching is one of the most common operations people perform with a GIS. In fact, addresses are probably the most commonly used form of geographic data. You'll find many uses for address matching. You could locate the address of a new restaurant that got rave reviews in the paper. You could find the location of your house in relation to flood zones. You might want to see the location of a house for sale listed in the newspaper, and see whether it's farther away from the flood zones. On a map you create for your friends, you could pinpoint the location of the party you're having to celebrate buying the new, flood-free house.

CHAPTER 4

Telling stories with digital maps

YOU'RE VISITING ITALY. Specifically, you're dragging a heavy suitcase through the alleyways of Venice, and you're lost—not surprising, since Venice is one of the most confusing cities in the world. You're also desperate. If you don't get to the *vaporetto* stop in the next five minutes, you'll miss your train and the plane home. Your Italian is nonexistent, and nobody seems to know a word of English. In desperation you drag out your Venice map—the one that looks like a plate of capellini—and a kindly shop owner traces out the route to the vaporetto with a blunt pencil. You make your plane on time.

While it's satisfying and simply fun to discover the wealth of information about a place like Washington, D.C., that's available from a digital map, don't forget that a map is at its core one of the best tools ever devised for communicating with other people: get lost in a strange city and a map will get you found.

Now that you've seen what a digital map—a GIS—can do, it's time to learn how to make your own maps. Not only is it easy, you'll also be surprised at the variety of situations in which a map is the best way to communicate information. You can use maps for school reports and reports at work, for directions to parties, garage sales, company picnics, family reunions, charity car washes. You can print them out or—unlike a paper map—easily ship them off by e-mail.

One of the first things you'll learn is that the appearance of a map makes a big difference. Each dot, line, or area on a map represents something in the real world—a city, a road, a country. You can draw them any way you want, but there are some traditions in map symbology that you probably already know without thinking: a double red line is usually a major highway, a tent is a recreation area, a tiny plane is an airport. Even colors are traditional: green means vegetation, blue means water. Needless to say, these symbols need to be consistent throughout the map.

If you keep these principles in mind, your maps will communicate effectively.

Exploration 7   A trip to Rio de Janeiro

At dinner one Sunday evening, your daughter tells you she needs a map of Rio de Janeiro for a school report due Monday morning. Together, you'll create a map that shows the streets, parks, and landmarks within a part of Rio known as Ipanema.

When finished, you'll print the map to include with the report she has also not started.

1   Start ArcExplorer, if necessary.

2 Click the Open Project button. Navigate to the *explore\rio* directory on the CD, choose the project file *rio.AEP,* and then click Open.

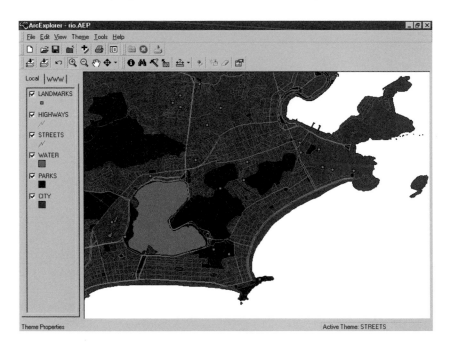

The map view contains themes of streets, highways, parks, landmarks, and water. However, the colors aren't very interesting, and it's hard to distinguish one theme from another. ArcExplorer assigns colors at random, so it's up to you to pick some that are more interesting.

First you'll change the CITY theme to light yellow so features in other themes will stand out.

3 Make the CITY theme active in the legend.

4 To display the Theme Properties dialog box, click the Theme Properties button or double-click on the theme name, CITY, in the legend.

In the Theme Properties dialog box, the theme name box shows you which theme you're working with. Classification Options shows you the way in which the features of this theme are displayed. In this case, the classification option is *Single Symbol,* which means that all the features in the theme are currently symbolized as shown in the dialog.

5 Click the Color box to display the Color dialog box. Choose *light yellow* and click OK.

6 Click OK in the Theme Properties dialog box to apply your change.

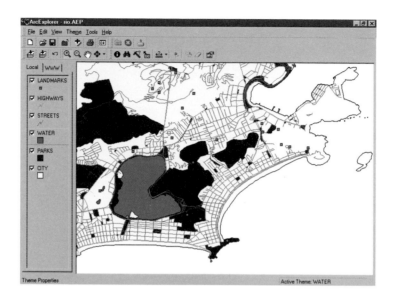

Do the same for the WATER theme.

7 Make the WATER theme active and use Theme Properties to make water *blue*.

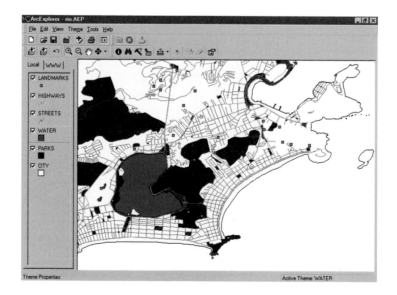

Now that water looks right, make parks green.

8 Make the PARKS theme active and use Theme Properties to make parks *green.*

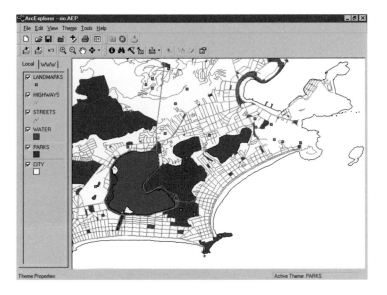

Now you'll select an appropriate symbol for streets.

9 Make the STREETS theme active and click the Theme Properties button.

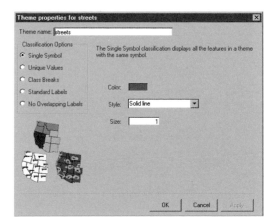

10 In the Theme Properties dialog box, click the Color box to display the Color dialog box. Choose *dark gray* and click OK.

Notice that the size property of the line is 1 (the default value). The size property controls the thickness of the lines on your map. The larger the value, the thicker the line. Keep a value of 1 for the STREETS theme.

11 Click OK in the Theme Properties dialog box to apply your change.

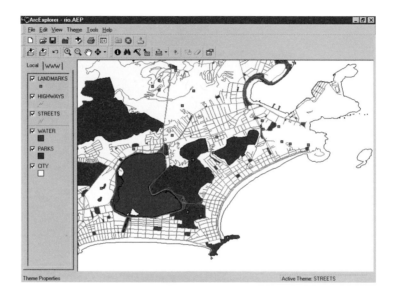

12 Make the HIGHWAYS theme active and use Theme Properties to make highways *red*. Since highways are usually bigger and used by more people than streets are, specify a size of *2*.

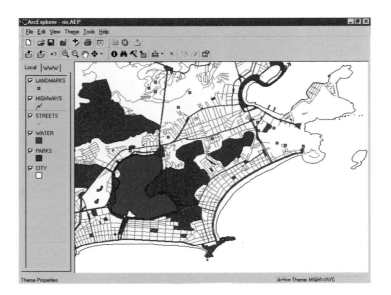

Now you'll change the landmarks symbol.

13 Make the LANDMARKS theme active, and use Theme Properties to make landmarks *red*. Use the Style pull-down menu to choose *Triangle marker*. Specify a size of *6*.

14 Click OK in the Theme Properties dialog box to make the changes.

Now zoom in on Ipanema, the most famous neighborhood of Rio de Janeiro.

 15 Click the Zoom In tool and create a box as shown in the following graphic.

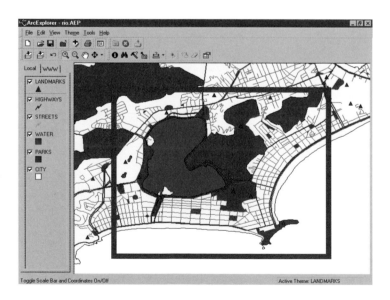

Now you'll create labels for each of the landmarks on your map so the teacher will know their names. You can do this in the Theme Properties dialog box also, using the Standard Labels and No Overlapping Labels buttons. The buttons create labels using a specified text font that you get from a field in the table.

Use Standard Labels to place the labels according to choices you set.

Use No Overlapping Labels to keep labels from crowding or overlapping. This is good for labeling features such as streets. You can also use this option to create a colored background under the label so the text is easily readable.

16 Click the Theme Properties button to display the Theme Properties dialog box.

17 Choose *No Overlapping Labels* under Classification Options.

Choose *NAME* as the text field.

In the label placement box, choose *Place on*.

Click on the *Mask labels* option.

Click in the *Mask color* box and choose *white*.

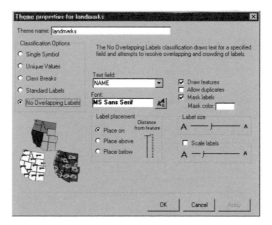

Click Apply to check the labels. They may appear too large on the map. If so, adjust the label size slider and then click Apply to check the size. Repeat this procedure until your labels are big enough to be legible, but not so big that they crowd each other out.

Click OK in the Theme Properties dialog box when you're finished.

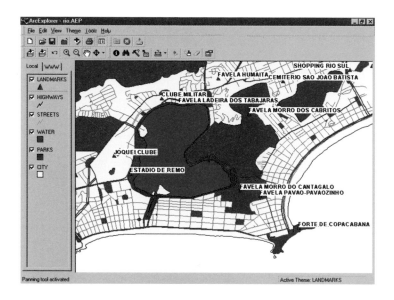

Since the white area in this map view is actually water, set the background of the map to blue.

18 Select Map Display Properties from the View menu. The Map Display Properties dialog box displays.

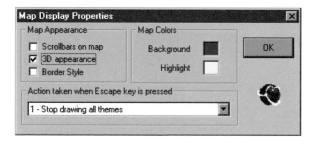

 19 Under Map Colors, click *Background* to display the Color dialog box. Choose *blue,* then click OK to close the dialog box.

20 Click OK to close the Map Display Properties dialog box.

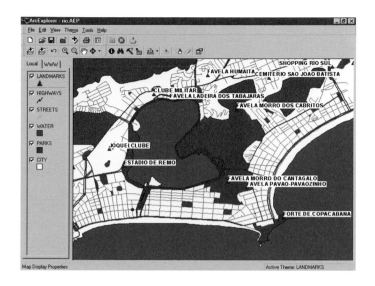

Now add a scale bar so the class can see how large this area is.

21 Choose Display Scale Bar from the View menu. A scale bar appears below the map view.

22 Right-click the scale bar and set the map units to *Decimal Degrees.* Set the scale units to *Miles* and the screen units to *Inches.*

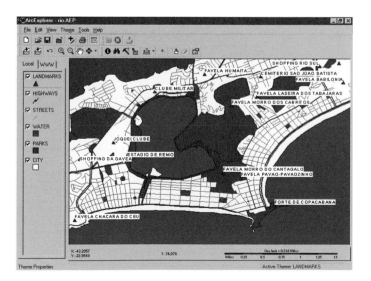

Your map is now ready to print.

23 Click the Print tool. The Print Map dialog box displays. This is where you choose a printer and where you give your map a title.

> **Print Map** ✕
>
> Map Title
>
> A map of Rio de Janeiro
>
> Printer
>
> My Color Printer ▾ 📋
>
> Print
>
> Cancel

24 In the Print Map dialog box, enter **A map of Rio de Janeiro** as the title of your map.

25 Click Print.

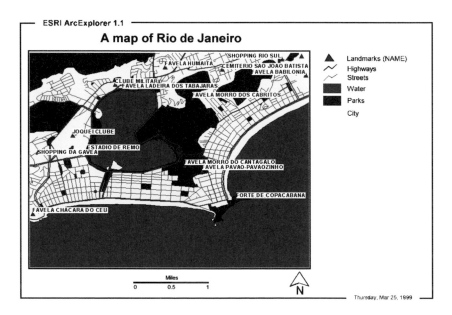

Your finished map will appear as shown here, formatted with north arrow, scale bar, legend, and title.

26 Close the project without saving any changes.

Exploration 8   Symbolize a map of Prague based on attributes

In the previous exploration, you symbolized all the features in a theme with the same symbol, but you can also use symbols that reveal more information about a feature. The decisions here aren't as simple as assigning green to parks and red lines to highways. For example, a theme of roads may have an attribute that describes the type of road—two-lane highway, four-lane highway, major street, and minor street. You can use this attribute to assign different symbols to each type so your audience can easily tell them apart.

Suppose you work for a university that offers its students a chance to attend summer school in the historic city of Prague in the Czech Republic. During the summer, students attend class Monday through Thursday and have the remaining three days to enjoy the sights and atmosphere of the city. To help recruit students, you want to create a map showing the historic center of Prague, its major tourist attractions, and restaurants. Then you want to save the map as an image that you can put on your university's Web page and in a printed student newsletter.

ArcExplorer

1 Start ArcExplorer, if necessary.

2 Click the Open Project button. Navigate to the *explore\prague* directory on the CD, choose the project file *prague.AEP,* and then click Open.

When the project opens, you see the historic area of Prague. Themes include parcels, streets, restaurants, major attractions, and water. Although each theme uses a single symbol to represent its features, each feature contains several different attributes. For example, the RESTAURANTS theme contains an attribute to identify whether each symbol represents a pub, a coffeehouse, or a full-serve, sit-down restaurant.

3 Make the PARCELS theme active. Use the Identify tool to click on some of the areas.

Notice that each area you click on has a land use attribute that specifies whether that piece of land is parkland, built-up area, pavement, or some other type. You may need to zoom in to distinguish one parcel from another.

4 Dismiss the Identify Results dialog box by clicking the X in its corner.

You'll assign different colors to each type of land use.

5 Click the Theme Properties button to display the Theme Properties dialog box.

6 To apply a different color to each unique land use, choose *Unique Values* under the Classification Options heading.

Choose *LAND_USE* as the field in the Field pull-down menu.

These default colors are probably not all appropriate for the features they represent, since the software assigns them at random. You'll choose some new ones.

In the Theme Properties dialog box, click on the color to the left of the words *built-up area*. The Symbol Properties dialog box displays. Choose *light red* as the color. Click OK.

Repeat this procedure for the remaining colors, clicking on each in turn and changing its color. You may not have to change all of them.

For *greenery,* assign light green.

For *land under cultivation,* assign green.

For *other,* assign light gray.

For *railway,* assign purple.

For *recreation,* assign red.

For *street, pavement,* assign dark gray.

For *water bodies,* assign blue.

When you're finished symbolizing land uses, check the *Remove outline?* box and then click OK in the Theme Properties dialog box.

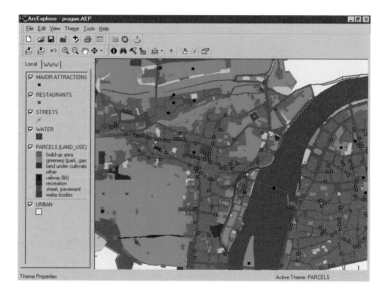

Readers of your map can now easily distinguish the different types of land use.

Now look at the attributes for the restaurants.

7 Make the RESTAURANTS theme active. Use the Identify tool to click on several of the points.

Each restaurant has a name and a type.

```
Identify Results                      [x]
Location: X: -744,584.8971  Y: -1,042,48
1 feature found
Feature:
KAJETANKA                            [▼]

Attributes:
ADDRESS = Hradcanske nam. 10
FeatureId = 220
ID = 612
NAME = KAJETANKA
TYPE = Cafe, Coffeehouse




Theme:  Restaurants
Shape Type:  Point
```

Next, you'll symbolize the restaurants with different point symbols so students can tell whether each is a sit-down establishment, a pub, or a coffeehouse. The procedure is much the same as the one you just followed for land uses.

8 After dismissing the Identify Results dialog box, make the RESTAURANTS theme active in the legend.

9 Click the Theme Properties button to display the Theme Properties dialog box.

Choose *Unique Values* under Classification Options.

Choose *TYPE* as the field in the Field pull-down menu.

You'll now choose different symbols for each type of restaurant.

In the Theme Properties dialog box, click on the symbol to the left of the words *Cafe, Coffeehouse.* The Symbol Properties dialog box displays.

Choose *orange* as the color, *Circle marker* as the style, and *6* as the size.

Repeat for the remaining symbols, clicking on each symbol in turn and modifying the symbol properties for each.

For Pub, choose *dark blue* as the color, *Triangle marker* as the style, and *6* as the size.

For Restaurant, choose *dark green* as the color, *Square marker* as the style, and *6* as the size.

When you finish symbolizing the restaurants, click OK in the Theme Properties dialog box.

Now that the different types of restaurants can be distinguished, symbolize and label the major attractions.

10 Make the MAJOR ATTRACTIONS theme active and use Theme Properties to make the attractions yellow. Use the Style pull-down menu to choose *Triangle marker.* Specify *6* as the size.

Choose *No Overlapping Labels* under Classification Options.

Choose *NAME* as the text field.

Click on the *Mask labels* option.

Click in the *Mask color* box and choose *white.*

Click Apply in the Theme Properties dialog box to view the labels in your map view. Adjust the label size with the label size slider if they appear too large. Click OK when you're satisfied with the label size.

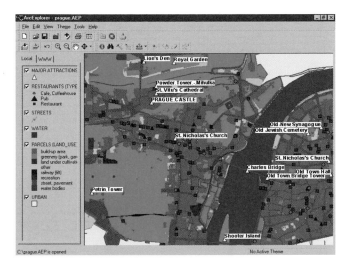

Your map is now well symbolized and labeled. All that's left to do is save your map so others can use it.

From the Edit menu, there are several options for saving your map view. Choose Copy to Clipboard if you plan to paste it into another Windows® program. Choose Copy to File if you want to create a file to use anytime.

Choose BMP (Windows bitmap) if your map view contains a scanned map or aerial photograph. Otherwise, choose EMF (enhanced metafile).

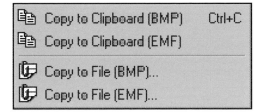

11 From the Edit menu, choose Copy to File (EMF). ArcExplorer prompts you to specify a name and a location for the new file. You might save the file in the directory *c:\prague* as *prague.emf.*

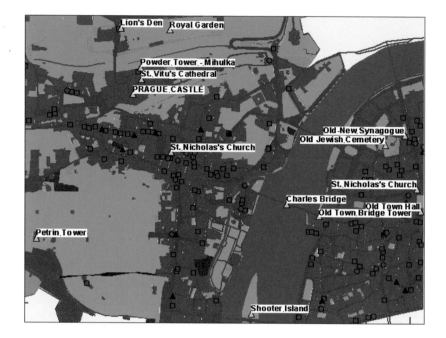

Now you can use the file in another Windows application such as Microsoft Word or in a drawing program like Adobe® Illustrator®. Your drawing program may allow you to convert the file to GIF or JPG format for use on your personal Web page.

12 Close the project without saving any changes.

You now have an informative map of Prague that gives students a good idea not only of the city's geography, but also its history and its recreational possibilities. You can put the map on the school's Web page, or include it in a newsletter mailed to students. Either way, you should have students banging at the door to attend the summer program.

Exploration 9   Share your map of New York City

Your boss has called a meeting to kick off a marketing campaign in New York. Your task is to come up with some maps that show relevant marketing data about the city; specifically, where the wealthier, more highly educated people (generally considered a demographically desirable group) live. You'll create an ArcExplorer project that your colleagues can take with them to study.

Since you can't create your own project on the CD, you need to copy the data from the CD to your local hard disk.

1   Create a folder on your PC to store the data. For example, you might create a folder named *gis_data* on your C: drive.

2   Copy the *newyork* folder from the *explore* folder on the CD to the location you just created.

Now that the data is on your PC, you're ready to create your maps.

3 Start ArcExplorer, if necessary.

4 Click the Open Project button. Navigate to the *newyork* folder you copied, choose the project file *newyork.AEP,* and then click Open. The project opens, and you see a map view of some residential areas of Manhattan. It contains several themes, including STREETS, WATER, PARKS, and CENSUS.

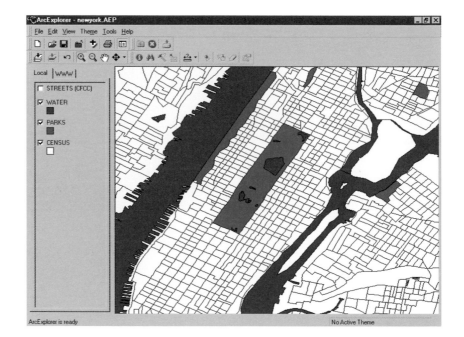

5 Make the CENSUS theme active and click the Identify button. Using your mouse, click around in several different areas to learn about them. The CENSUS theme contains the same kind of demographic information about Manhattan that you saw in the Washington, D.C., exploration.

```
Identify Results                          ☒
Location: X: -73.9509  Y: 40.7808
1 feature found
Feature:
┌────────────────────────────────┬──┐
│150                             │▼ │
└────────────────────────────────┴──┘
Attributes:
┌──────────────────────────────┬─┐
│ABLEENGL = 278               │▲│
│AGE05_09 = 52                │ │
│AGE10_14 = 35                │ │
│AGE15_17 = 31                │ │
│AGE15_19 = 50                │ │
│AGE18_19 = 19                │ │
│AGE20_24 = 170               │ │
│AGE21_24 = 163               │ │
│AGE25_34 = 699               │ │
│AGE35_44 = 547               │ │
│AGE45_54 = 250               │ │
│AGE55_64 = 150               │ │
│AGE65_74 = 77                │ │
│AGE75_84 = 45                │▼│
└──────────────────────────────┴─┘

Theme:  census
Shape Type:  Polygon
```

6 Close the Identify Results dialog box.

Now you'll create a couple of maps so you can see how some of these attributes are distributed throughout Manhattan. Specifically, you'll be looking for areas that have a high proportion of high-income wage earners, and areas that have a high proportion of college graduates. The areas where these groups intersect will be your target marketing areas.

First, you'll create a map of median income.

7 Click the Theme Properties button to display the Theme Properties dialog box.

Choose *Class Breaks* under Classification Options.

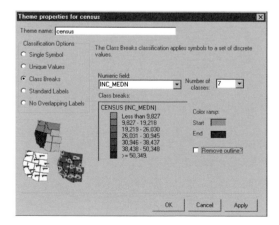

The Class Breaks option is used to create a graduated color map, such as this one, from numeric data. Similar numeric values are grouped together as ranges, or classes. A different color is applied to each range.

8 In the *Numeric field* pull-down menu, scroll down and choose the field *INC_MEDN*.

9 In the *Number of classes* pull-down menu, choose *7*.

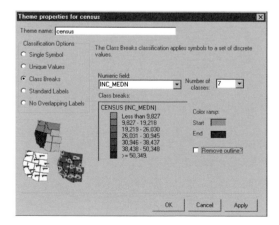

Seven is a good number because most people can visually distinguish about that many classes, give or take a couple. Depending on the quality of your monitor or printer, however, five classes may be a more practical limit.

Now you'll create a color ramp to represent the median income. A color ramp uses colors to indicate rank or order among classes. The colors progress from light to dark. With numeric data, lower values should use lighter colors and higher values should use darker colors. Different shades of the same or related colors work well; a progression of light red, red, and dark red is easier to interpret than green, blue, and red.

10   Click the Start color box to select a starting color for your color ramp. The Color dialog box will display.

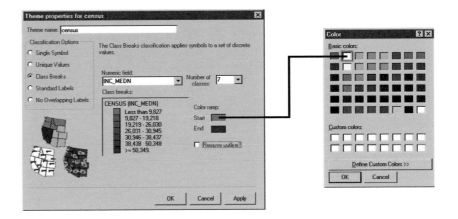

Select a shade of light yellow and then click OK.

11 Click the End color box to select the ending color for your color ramp. This time, select a shade of dark red and then click OK.

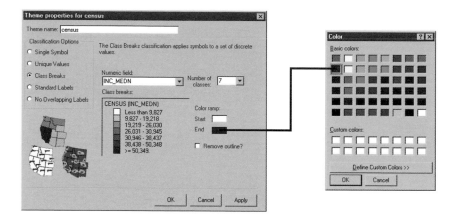

12 Click OK in the Theme Properties dialog box to make the map.

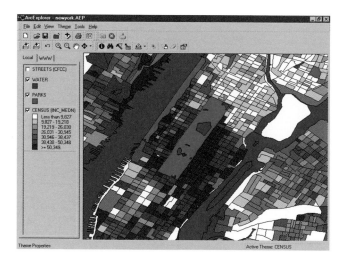

Your map shows lower income with lighter colors and higher income with darker colors. As you might expect, the higher-income groups tend to be clustered around Central Park. In fact, some very wealthy folks live in this area.

You can easily change the subject of the map by selecting a different demographic field—for example, the areas that have a high number of college graduates.

GIS for Everyone

13 Click the Theme Properties button to display the Theme Properties dialog box. In the *Numeric field* pull-down menu, scroll down and choose the field *COLGGRAD*.

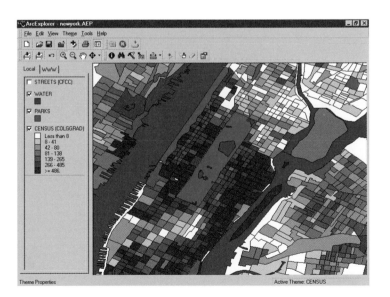

14 Click OK in the Theme Properties dialog box.

The subject of your map changes to show the distribution of people who graduated from college. Perhaps not surprisingly, it has the same general appearance as the previous map. You now have a pretty good idea of where your marketing campaign will be directed.

You'll save this map as an ArcExplorer project. That way, not only will you be able to view the map quickly at a later time, but you can also distribute your map digitally for others to see.

15 Choose Save As from the File menu. Specify a file name of *default.aep*. Save the project in the *newyork* folder you created earlier. Click Save. The project is saved to your local hard disk. (See the ArcExplorer Help topic "Zip archives" for important information on projects named *default.*)

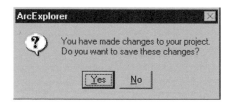

 16 Click the Theme Properties button to display the Theme Properties dialog box. Specify some different attributes in the *Numeric field* pull-down menu. You might also wish to experiment with different colors in the color ramp.

When you're done looking at other attributes, open the project you saved to display your college graduate map.

 17 Click the Open Project button. You'll be asked if you want to save changes to your project.

Click No. The Open ArcExplorer Project dialog box displays.

Click on the project *default.aep*. Click Open.

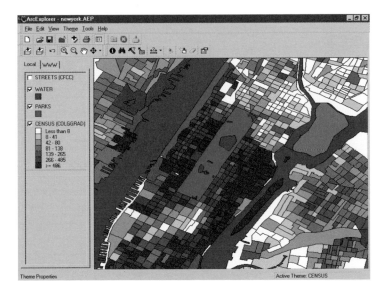

The map of college graduates in Manhattan once again displays on your screen. You can e-mail the map to your colleagues as a file attachment, and begin to plan your company's marketing campaign.

 18 Close the project without saving any changes.

SHARE A PROJECT

ArcExplorer projects that you create contain paths to folders that reside on your computer, such as *C:\gis_data\newyork*. Before you can share your ArcExplorer project with others, you must edit the project file so that all paths are relative. That way, the projects will open on all computers, regardless of the path where the data resides. See the ArcExplorer Help topic "Creating relative pathnames" to learn how to edit project files so they work on other computers.

CHAPTER 5

Building the digital map

UP TO THIS CHAPTER, you've been living a pretty sheltered life. We've provided you the appropriate data files for cities such as San Diego and Prague so you can get accustomed to working with a GIS without worrying about other issues. But the reality is that geographic data in the actual world comes in many different formats. This chapter introduces you to some of them, and then helps you put these diverse formats together to make an even more powerful map.

Varieties of geographic data

Geographic data is information about the earth's surface and the objects found on it. This information comes in three basic forms: map data, attribute data, and image data.

Map data contains the location and shape of geographic features. To represent real-world objects, maps use three basic shapes: points, lines, and areas (in a GIS, these are commonly referred to as points, lines, and polygons). Any object can be represented using one of these shapes.

Points represent objects that have discrete locations and are too small to be depicted as areas, such as schools, churches, train stations, and firehouses. Lines represent objects that have length but are too narrow to be depicted as polygons, such as freeways, roads, railroads, bridges, and creeks. Areas represent objects too large to be depicted as points or lines, such as large rivers, parks, lakes, golf courses, and forests.

Map Data

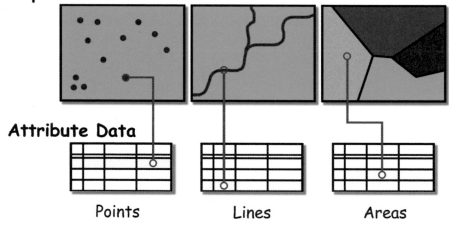

Attribute Data

Points   Lines   Areas

Attribute data is the descriptive data that GIS links to map features. Attribute data is collected and compiled for specific areas like states, census tracts, cities, and so on, and often comes packaged with map data.

Image data includes such diverse elements as satellite images, aerial photographs, and scanned data (data that's been converted from printed to digital format).

Geographic data comes in many different formats.

Varieties of geographic data files

The format of the data you choose must be compatible with your GIS and the data you already have. While formats can be a complex subject, at this point in your GIS career there are only a few that you need to know to begin creating good maps.

Shapefiles

One of the most common formats you're likely to encounter, and the easiest to use, is the shapefile. A shapefile can represent a point feature such as a city, a line feature such as a highway, or an area feature such as a county. Shapefiles store the geometric location and attribute information of these features.

Shapefiles are actually made up of at least three underlying files that must be kept together in the same folder or directory. Most of the time they're invisible to you and you don't have to worry about them. If you have to, you can identify them by their extensions (the suffix on the end of a file name): .shp, .shx, .dbf. You're most likely to encounter these file types when you unzip a compressed file.

One reason for knowing about these three is that you can use them as an alternative way to add a theme to an ArcExplorer map view. Although most of the time you'll add a theme by using the Add Theme button, the alternative method is to drag any of the three file types listed (SHP, SHX, DBF) from a Windows folder directly into the map view.

ArcInfo coverages

Another common format you're likely to run into is called an ArcInfo™ coverage.

ArcInfo is another kind of GIS software from ESRI, the company that makes ArcExplorer. Like shapefiles, coverages also store geographic features such as points, lines, and areas, along with their attributes. Some ArcInfo coverages can contain more than one type of feature. For example, a coverage containing area features such as land parcels may also contain line features that store information about the boundaries between the parcels.

Since a theme in ArcExplorer can only represent one class of features from an ArcInfo coverage, you choose which class you want the theme to represent. However, you can add several themes to a view from the same ArcInfo coverage, each based on a different class.

ArcInfo interchange files

Often, ArcInfo coverages are contained within files with a .e00 extension. These have to be converted to coverages using a special utility program called IMPORT71, which is included on the companion CD.

Instructions for installing and using IMPORT71 are found in appendix D.

Images

Images are intrinsically interesting. It would probably take about a thousand words to explain why. Instead, we'll just describe the two types of images you're most likely to run across, aerial (or satellite) photos and scanned images.

Aerial photographs

Aerial photographs make compelling backgrounds for your other GIS data. You can display the location of flood zones on an image of your hometown, or simply show a bird's-eye view of your neighborhood.

On the Internet, you're most likely to find aerial photographs in a format called Digital Orthophoto Quads, or DOQ. Most are free. There are also many private vendors that sell aerial photos and satellite imagery.

A DOQ image of Boston, Massachusetts, obtained for free via the Internet.

Scanned maps   Scanned maps, like aerial photos, can be used in the background with your other geographic data.

The type of scanned map you're most likely to encounter is the Digital Raster Graphic (DRG), scanned from a U.S. Geological Survey (USGS) standard-series topographic map. DRGs are increasingly being made available over the Internet for free. You can use them to plan your outdoor activities such as hiking or camping, or to find out how high that hill really is that you're planning to climb.

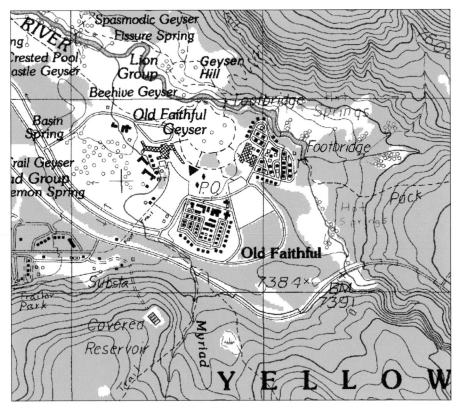

A DRG image showing part of Yellowstone National Park.

A great source of high-quality yet inexpensive scanned topographic maps is a product called Sure!MAPS RASTER from Horizons Technology, Inc. Complete coverage is offered for the United States. You'll learn how to obtain these maps in chapter 6.

Image formats   Images come in a variety of formats with their own extensions. You don't have to worry about how they work, but it's helpful to be aware of the different types so you can recognize them.

The most common format you're likely to encounter is TIFF (Tagged Image File Format). TIFF is a standard in the desktop-publishing world and serves as an interface to several scanners and graphic arts packages.

Other image formats and their extensions that ArcExplorer can read are: BMP and DIB files, which are Windows bitmap formats; GIS, LAN, BIL, BIP, BSQ, RLC, SUN, RS, and RAS.

You'll learn how to access links to sites offering free or inexpensive images in chapter 6.

Exploration 10 Make a map of Austin, Texas, from digital data

Photo courtesy of Abhijit Jas

You've done such a good job in New York that your boss has given you a promotion. You get to manage your firm's new satellite office in Austin, Texas. Now you have to find a place to live. You want a house that's not located in a flood zone, and you want to find out where all the schools and hospitals are.

You have several shapefiles of themes such as streets, highways, schools, hospitals, and streams. You have a nice aerial photo that will make an interesting background. And you have an ArcInfo coverage of hundred-year flood areas.

ArcExplorer™

1 Start ArcExplorer, if necessary.

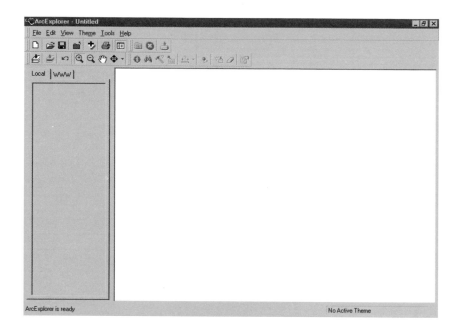

You see a blank map view and a blank legend.

2 Click the Add Theme button. The Add Theme(s) dialog box displays.

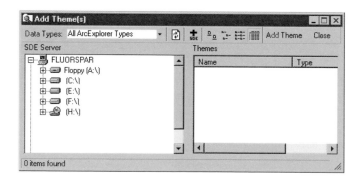

In this box are two windows. On the left is a list of all the drives and directories on your computer. This is where you find your data.

3 On the left side of the dialog box, find your CD drive. Then go to the *explore\austin* directory on the CD.

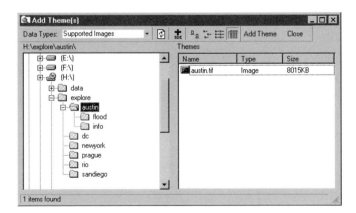

When you open the *austin* folder, you'll see several different themes, in different formats, listed in the window at the right. At the top of the box, in the Data Types window, *All ArcExplorer Types* is the default choice. This means the right window will show shapefiles, coverages, or image files. If you have a lot of different data types available, you can use the Data Types pull-down menu to limit the kinds of file formats that will be listed in the window.

Since aerial photographs work well as backgrounds, you'll look to see what images are available.

4 Using the Data Types pull-down menu, choose *Supported Images*.

You see that only one image, called austin.tif, is listed in the themes list. You'll add that image to the map view.

5 Click on the name of the file austin.tif. Click Add Theme, then click Close.

The theme is added to the legend as AUSTIN (Image).

6 Turn on the theme AUSTIN in the legend.

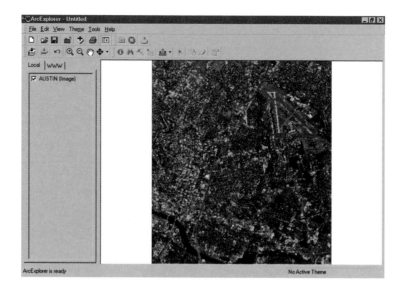

Now you'll add your shapefiles to the map view.

7 Click the Add Theme button. In the Add Theme(s) dialog box, choose *Shapefiles* as the data type.

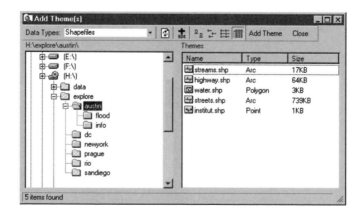

Only shapefiles appear in the Themes list.

8 Click on water.shp and Add Theme, streams.shp and Add Theme, streets.shp and Add Theme, highway.shp and Add Theme, institut.shp and Add Theme.

9 Turn on the five new themes in the legend.

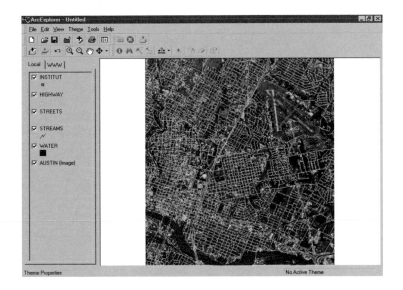

Since ArcExplorer assigns random colors to the new themes, you may need to assign more appropriate ones.

10 Use Theme Properties in turn for each theme to assign appropriate colors and symbols (you learned how to do this in chapter 4). Following are some suggestions:

WATER: *blue*

STREAMS: *blue*

STREETS: *light gray* with a size of *1*

HIGHWAYS: *red* with a size of *2*

INSTITUT: Use a unique value classification and choose *TYPE* as the field. For each type, assign symbols as follows: for Cemetery, use *green* as the color, *Triangle marker* as the style, and *7* as the size; for School, use *yellow* as the color, *Circle marker* as the style, and *7* as the size; for Hospital, use *red* as the color, *Square marker* as the style, and *7* as the size.

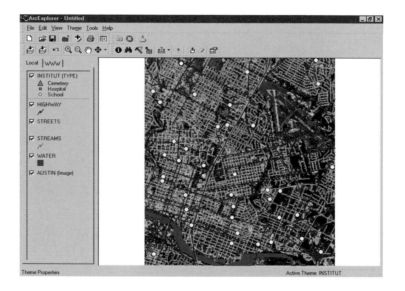

Your map is looking pretty good. All that remains is to add the coverage of flood areas to the map view.

11   Click the Add Theme button. Choose *ArcInfo Coverages* from the Data Types pull-down menu.

In the Add Theme(s) dialog box, navigate to the *flood* folder. Here you'll see a theme named pat.adf.

Click on the theme and then click Add Theme. The theme is added to the top of your legend as FLOOD.PAT.

12   Turn on the FLOOD.PAT theme.

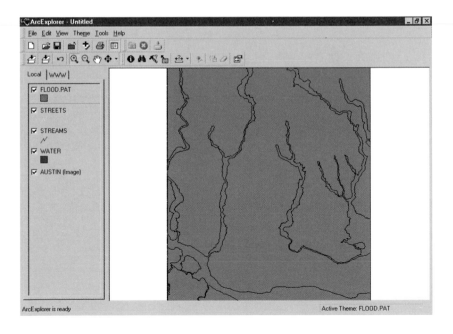

ArcExplorer displays themes in the order they appear in the legend, from bottom to top. Since the flood theme is on top of the list, it covers up all the other themes. You'll move it down the list a bit.

13   Click on the name of the FLOOD.PAT theme in the legend and, holding down the mouse button, drag the mouse pointer down until it's between the AUSTIN (Image) and WATER themes, then release the mouse button.

Now you'll need to symbolize the FLOOD.PAT theme to show areas within a Special Flood Hazard Area (SFHA). These are areas expected to flood at least once within the next hundred years and are commonly called hundred-year flood zones.

Make the FLOOD.PAT theme active.

 Click the Theme Properties button.

In the Theme Properties dialog box, select *Unique Values* under Classification Options.

Select *SFHA* as the field.

In the *Discrete values and symbols* section of the dialog box, you see that ArcExplorer automatically assigned random colors to each unique classification IN and OUT.

IN—Inside a hundred-year flood zone. Symbolize these areas in the map view as *green*.

OUT—Outside a hundred-year flood zone. Use a transparent symbol for this.

14 Under *Discrete values and symbols,* click on the color box next to IN.

In the Symbol Properties dialog box, select *light green* for both the color and outline color, and *Dark gray fill* as the style. Click OK.

15 Under *Discrete values and symbols,* click on the color box next to OUT.

In the Symbol Properties dialog box, select *Transparent fill* as the style. Click OK. You needn't choose a color when the style is *Transparent fill.*

16   Click Remove Outline in the Theme Properties dialog box. Click OK.

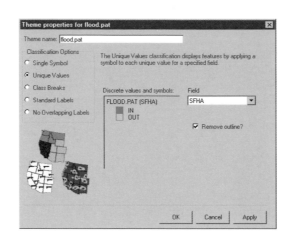

Your map view looks great. You know where you should be if you're ever in Austin and it starts to rain heavily. More importantly, you're prepared to use many different data types together if you ever need to. Now go ahead and explore your map. Use the pan and zoom tools to get a closer look, or use Identify and MapTips to see the names of streets, highways, streams, schools, and hospitals.

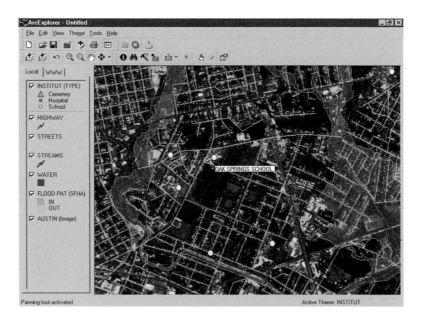

CHAPTER 6

Bringing the world into your digital map

LEARNING GIS is a little like learning to fly. Like flying, there are some basic principles to learn, as well as a new vocabulary—terms like "shapefile" and "map units." Like flying, it takes practice to get good at it. Like flying, it gives you a whole new way of looking at things.

And like flying, GIS is essentially just a means for you to get out there and discover the world around you on your own.

If you've gotten this far, you're about ready to solo. This last chapter shows you some places to begin bringing the world around you into your own digital maps, and then gives you some ideas about places to go after that.

As Internet use has exploded, so has the amount of geographic data available on the Internet. Once you start exploring it, you'll be astonished by its quantity and its diversity.

GIS for Everyone companion CD

The *GIS for Everyone* CD–ROM features more than 500 megabytes of digital geographic data for you to explore. The data covers topics such as population, agriculture, income, economics, health, geological hazards (volcanoes and fault lines), and geographic features such as roads, rivers, lakes, cities, forests, and mountain peaks.

Which files are used and how they're presented in a map is up to you. The result of these creative efforts might be something like the image shown below, made up of eleven different map layers. The order of the layers (from bottom to top), their colors, and the symbols used to represent each were selected after each layer was added. The final map image is in the hands of the mapmaker—you.

All the geographic data on the CD is located in the *data* and *explore* folders. These geographic files are listed and described in the CD's data dictionary in the file *datadict.htm*. Each entry, organized to reflect the data directory structure, briefly describes the file's subject matter, geographic coverage, and date (if appropriate). Each entry also has a link to display information about all the attributes associated with each theme.

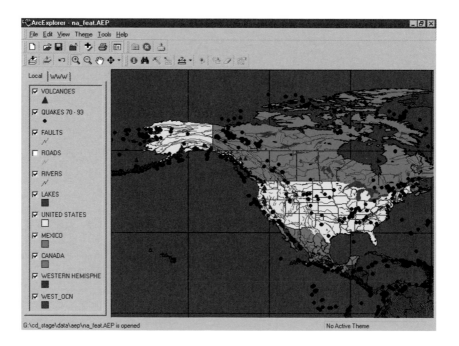

Exploration 11   View data on the CD

We've included a couple of ArcExplorer projects to get you started exploring the data on the CD. These projects represent only a fraction of the possible things you can explore with the data.

1 Start ArcExplorer, if necessary.

2 Click the Open Project button. Navigate to the *data\aep* directory on the CD.

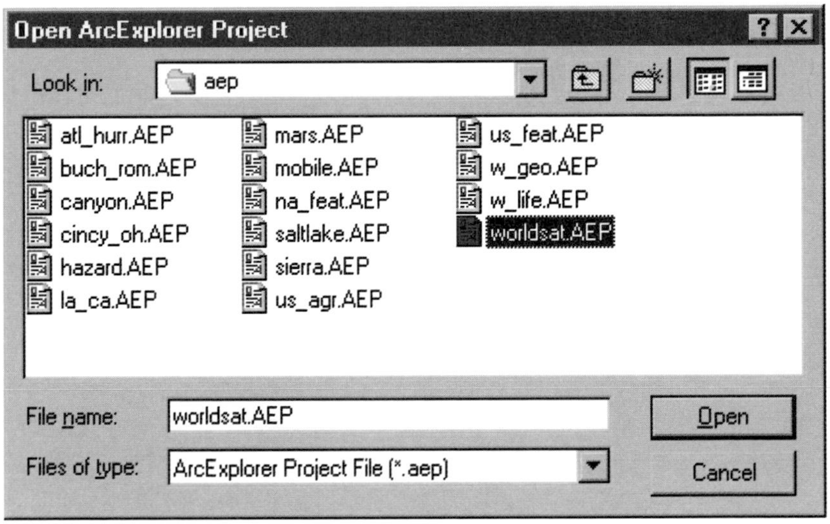

Open ArcExplorer Project

Look in: ⬜ aep

📄 atl_hurr.AEP	📄 mars.AEP	📄 us_feat.AEP
📄 buch_rom.AEP	📄 mobile.AEP	📄 w_geo.AEP
📄 canyon.AEP	📄 na_feat.AEP	📄 w_life.AEP
📄 cincy_oh.AEP	📄 saltlake.AEP	📄 worldsat.AEP
📄 hazard.AEP	📄 sierra.AEP	
📄 la_ca.AEP	📄 us_agr.AEP	

File name: worldsat.AEP

Files of type: ArcExplorer Project File (*.aep)

Open
Cancel

In the Open ArcExplorer Project dialog box, you see a lot of sample projects created from the data on the CD. Take a look at some of them.

3 Click on any of the .AEP files. The following table describes the available project files:

Project	Description
na_feat.AEP	Explore North American natural and man-made features such as volcanoes, earthquake sites, roads, rivers, lakes, and state boundaries.
us_agr.AEP	Explore farm statistics for the United States. Average Farm Size by County is the default.
us_feat.AEP	Explore natural and man-made features such as forests, mountains, rivers, lakes, highways, and major cities in the United States. See where the Continental Divide is.
w_geo.AEP	Explore world features such as volcanoes and fault lines. See the locations of the twelve major tectonic plates.
w_life.AEP	Explore demographic data for all the countries of the world.
mars.AEP	Explore an image of the surface of Mars.
buch_rom.AEP	Explore land use and demographic information about Bucharest, Romania.
cincy_oh.AEP	Explore an Aerotopia aerial photograph of downtown Cincinnati, Ohio.
la_ca.AEP	Explore building footprints, including heights.
mobile.AEP	Explore the predicted storm surge resulting from five categories of hurricanes landfalling near Mobile, Alabama.
worldsat.AEP	Explore a satellite image of India and southern Asia, including the Himalayas.
sierra.AEP	Explore an HTI Sure!MAPS RASTER 1:250,000-scale topographic map of the Sierra Nevada Mountains, including 14,494-foot-high Mount Whitney.
canyon.AEP	Explore an HTI Sure!MAPS RASTER 1:100,000-scale topographic map of the Grand Canyon.
saltlake.AEP	Explore an HTI Sure!MAPS RASTER 1:24,000-scale topographic map of Salt Lake City, site of the 2002 Winter Olympic Games.
ranier.AEP	Explore an HTI Sure!MAPS RASTER 1:100,000-scale topographic map of Mount Rainier National Park, Washington.

These are just a few of the many projects you can create on your own using this data.

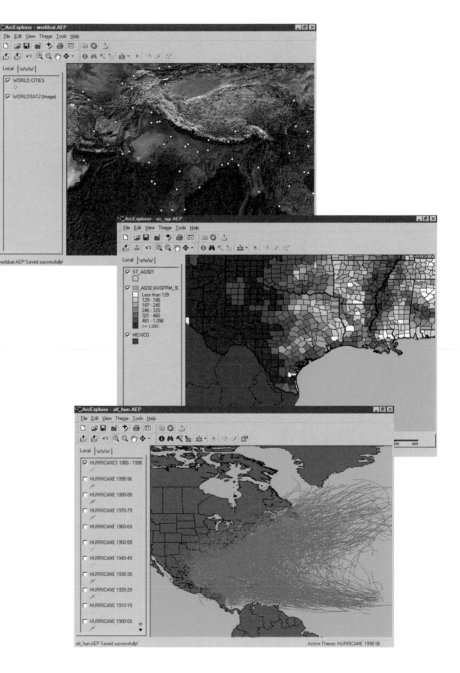

GIS for Everyone companion Web site

The *GIS for Everyone* Web site is designed to help you find geographic data and learn more about GIS. You'll find the site at **www.esri.com/gisforeveryone**.

Here are some of the things you'll find at the site:

Learn the basics of GIS

If you enjoyed creating maps in the explorations and want to learn more about GIS and all it has to offer, this section is for you. It includes numerous educational materials to help you learn about this field. You can read how GIS is making a difference in communities around the world. And you'll learn how you can make GIS a career.

Download geographic data for your neighborhood

Use the special access code inside the back cover of this book to download data for your neighborhood. We'll help you through this process in exploration 12.

Find geographic data on the Web

Here's your entry point to the vast amount of geographic data on the Web. You can start by exploring some of the best and most interesting Web sites that offer geographic data you can download and use in ArcExplorer. You'll find a searchable, organized list of sites offering links to data, data clearing-houses, individual Web sites, search engines, and commercial vendors.

Get software

Here's where you can download next-generation versions of ArcExplorer so you'll always have access to the latest in functionality. You'll also find utilities, such as import and conversion programs, that enhance your use of geographic data.

Connect with other readers

This is where you can contact other readers and ArcExplorer users in a discussion forum devoted to this book. You might post a question asking about a certain type of data, or you may just wish to share a unique application of ArcExplorer that you discovered.

ArcData Online

ArcData Online is an ESRI Web site that lets you browse through a wide collection of GIS data sets, make maps on demand, and download selected data. Most of the data you download contains projects that you can simply drag and drop into ArcExplorer. The Web address for ArcData Online is **data.esri.com**.

ArcData Online works as follows: First, you access the ArcData Online site with your Web browser. Then you create preview maps online for different geographic data types. The preview map images you create with ArcData Online in your Web browser are yours to use freely in any application or publication. The data displayed can also be downloaded in shapefile format for use in ArcExplorer or other GIS software. Much of the data available through ArcData Online may be downloaded for free. Other data sets may be licensed and downloaded at low cost.

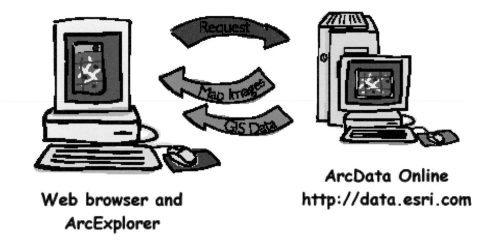

Web browser and ArcExplorer

ArcData Online http://data.esri.com

ArcData Online is evolving into a portal for geographic data. It will include an advanced search capability to find geographic data by location, subject, or type. The site will also include information on hundreds of data sets available from ArcData Online and elsewhere on the Web.

Following is a summary of the selection of data available on ArcData Online.

Free data sets   The site offers a vast amount of geographic data that may be downloaded and licensed free of charge. New data is added periodically. All of this data is derived from a selection of commercial data and includes both basemap and thematic data at the U.S. and world levels.

Basemap data contains geographic features used for locational reference, such as roads, railroads, rivers, lakes, political boundaries, buildings, and airports. A road atlas is a basemap, as are topographic maps.

Thematic data contains information that shows the distribution of a specific attribute of some natural or man-made phenomenon. A map showing population distribution by county is a thematic map.

U.S. basemap data   An excellent source of publicly available local geographic data is the U.S. Census Bureau's TIGER/Line® files. (TIGER is an acronym for Topologically Integrated Geographic Encoding and Referencing system.) This data is often enhanced and sold by commercial companies.

TIGER/Line files contain line, boundary, and landmark features. Lines include roads, railroads, rivers, and utility lines. Boundaries include census tracts and blocks, counties, congressional districts, and school districts. Landmarks include schools, churches, parks, cemeteries, apartment buildings, and factories.

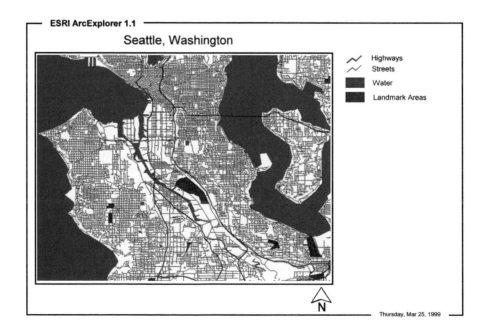

ESRI ArcExplorer 1.1

Seattle, Washington

Highways
Streets
Water
Landmark Areas

N

Thursday, Mar 25, 1999

World basemap data World basemap data allows you to create maps of geographic features such as country boundaries, rivers, roads, railroads, and airports.

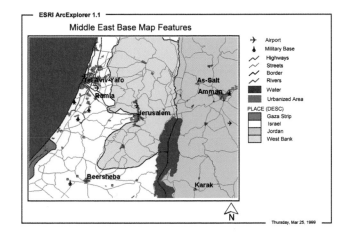

The ArcData Online collection of world basemap data includes data layers from a variety of ESRI products, including ArcWorld™, ArcAtlas™, Digital Chart of the World, and Data and Maps. ESRI has assembled these products into a single database.

To download data, you first select an area of interest. Then you preview your data in your Web browser. There are eleven display levels, ranging from a small-scale global display to a medium-scale regional display. The more you zoom in on an area, the more layers of data there are available to download. Here you see data for Europe being previewed online prior to download.

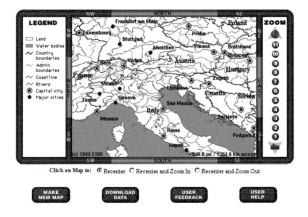

U.S. thematic data   Here you can access thematic data to create maps of information about the United States and Americans.

This database allows you to create thematic maps of selected census data and FBI crime data at the state and county level. The census data includes information on population, age, housing, households, and educational attainment.

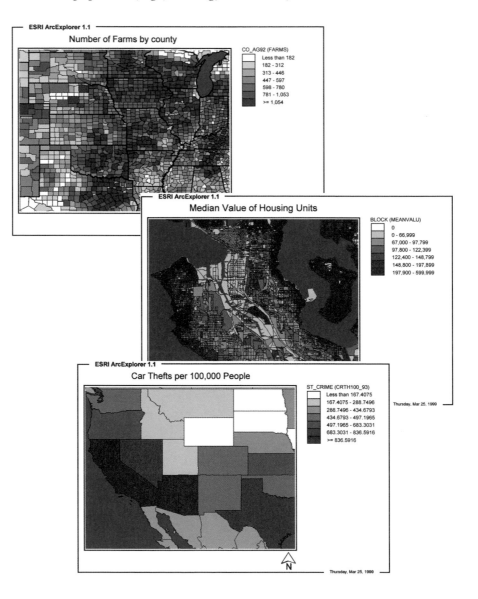

World thematic data

World thematic map data includes data layers from ESRI's ArcAtlas: Our Earth. You can download data to create maps of information about the earth, including its people, plants, and environment. Here you see a map showing yearly rainfall amounts in South America.

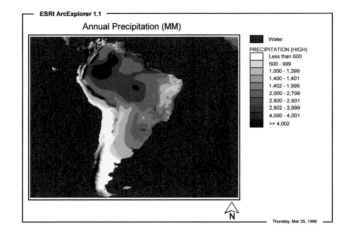

Accessing the data is simple. First, you select a topic such as elevation or population density. Then you choose an area of interest. You then preview your data and make adjustments to the size of the area you want to download. Finally, you download the data. Changing the topic is easy. Here you see four different topics previewed online.

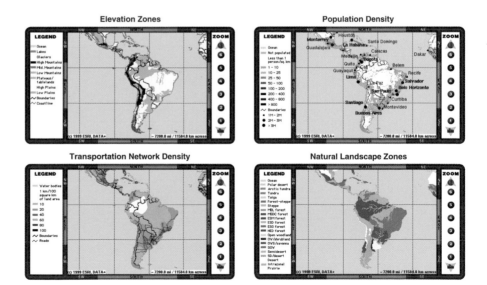

Commercial data sets
The site includes several commercial data sets. You may find what you need for free elsewhere, but when you can't, these low-cost data sets can often give you what you need, and may be more up to date and accurate.

GDT Dynamap/2000
Dynamap/2000® from Geographic Data Technology, Inc., is an enhanced version of the TIGER data. GDT has updated many of the TIGER data layers to add new geographic features and improve accuracy. The data you explored for Washington, D.C., is from GDT.

The Dynamap/2000 database contains more than fourteen million addressed street segments along with postal boundaries, landmarks, and water features. It offers the most comprehensive U.S. street and address data available today. The Dynamap/2000 data available through ArcData Online is intended for users interested in relatively small geographic areas. Users can download data by ZIP Code area.

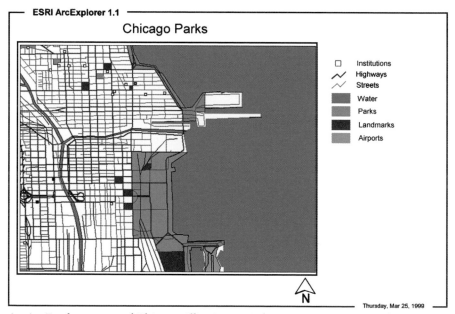

An ArcExplorer map of Chicago, Illinois, created using Dynamap/2000 data. Copyright © Geographic Data Technology, Inc.

HTI Sure!MAPS RASTER

Sure!MAPS RASTER maps from Horizons Technology, Inc., are scanned, full-color, U.S. Geological Survey topographic maps at 1:24,000, 1:100,000, and 1:250,000 scales. Topographic maps represent a three-dimensional surface on a flat piece of paper. These are the types of maps commonly used for outdoor recreation—for planning hikes and camping trips—as well as urban planning, resource development, and surveying. The map of San Diego you first looked at in chapter 2 is an HTI Sure!MAPS RASTER map.

The cultural, topographic, and environmental information contained on topographic maps can enhance the usefulness and usability of other data sets. Each map contains colorful details including topographic contours, land use features, political boundaries, streets, buildings, landmarks, and cultural information. These maps can be downloaded for user-defined areas anywhere in the United States.

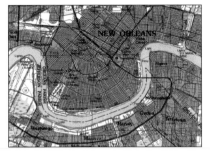

New Orleans, Louisiana, 1:100,000 scale. Copyright © Horizons Technology, Inc.

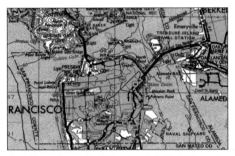

San Francisco, California, 1:250,000 scale. Copyright © Horizons Technology, Inc.

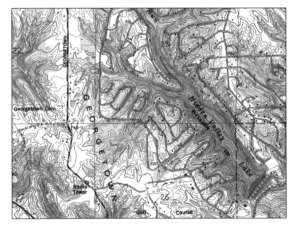

An area near Lawrenceburg, Indiana, 1:24,000 scale. Copyright © Horizons Technology, Inc.

FEMA flood data   Through ArcData Online, you can download flood data for a selected U.S. county or determine the flood risk at a specific location. This data is commonly used by insurance companies and other organizations to determine the flood risk to buildings and properties. You completed an exploration in chapter 5 using this type of data for Austin, Texas.

With ArcExplorer, you can display flood boundaries with other themes such as streets or neighborhoods to display zones of potential flood risk. The ArcExplorer map below shows flood data for the area near the city of Harrison in southwestern Ohio.

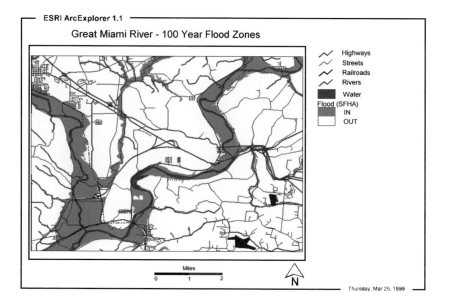

Using ArcData Online, you can also enter an address and see an image of flood risk there, right in your Web browser, without downloading data.

U.S. environmental hazards

VISTA Information Solutions' Environmental Geographics contains location and attribute data for more than 3.5 million hazardous waste generators, both regulated and unregulated. These include leaking tank sites, toxic spills, and other sites affecting the environment throughout the United States. The Environmental Geographics data available through ArcData Online is provided in compressed shapefile format for selected ZIP Code areas.

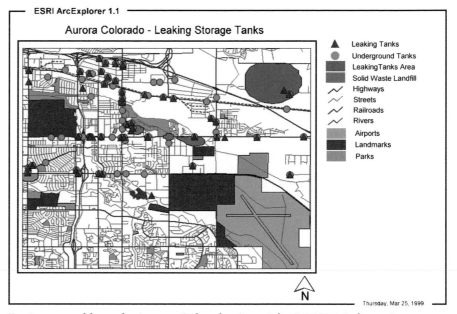

Environmental hazards, Aurora, Colorado. Copyright © VISTA Information Solutions, Inc., Geographic Data Technology, Inc.

*AND Mapping
European road atlas*

This database includes a road network as well as country, province, and city boundaries; major lakes and rivers; and railways. It offers the most comprehensive European basemap data available today.

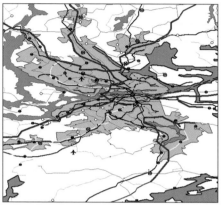

An image of Stockholm, Sweden, saved from ArcExplorer. Copyright © AND Mapping.

*TeleAtlas European
city data*

Here you can access the TeleAtlas street database for Europe, known as StreetMap. The StreetMap database contains highways, streets, railroads, postal areas, parks, and water features for the larger cities in Europe. It offers the most comprehensive European street data available today.

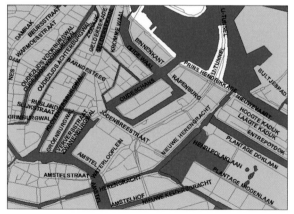

An image of Amsterdam, The Netherlands, saved from ArcExplorer. Copyright © TeleAtlas.

**A place to check
often**

ArcData Online is updated continually as new data sets become available for publishing and distribution. Future types of available data may include satellite imagery and aerial photos.

Exploration 12 Download data for your area

As a bonus for buying this book, you can access data for your own neighborhood, or for any area in the United States, using the special access code inside the back cover. You'll receive up to three themes for your selected area, depending upon availability, including GDT streets data, FEMA flood data, and census data.

The data will be provided to you as a compressed file you can download to your local computer and then drag and drop into ArcExplorer.

1 Using your Web browser, access the *GIS for Everyone* Web site at **www.esri.com/gisforeveryone**.

2 Access the Download Data for your Neighborhood link to display the download form.

3 Enter your name, the twelve-digit access code inside the back cover of this book, and the ZIP Code for which you want to download data. Be sure you enter the ZIP Code digits accurately; you can only do this once.

If you're outside the United States or don't have a specific ZIP Code you wish to download, several sample data sets are available to download from this page.

4 Click the Download button to process your request. It may take a few moments for your data file to be created.

5 When your file is ready, you'll be given a link from which you can download a compressed file containing the available themes of streets, flood zones, and census statistics.

Download the file and save it in a directory you've created on your PC (for example, *C:\homedata*).

Now you can add your themes to ArcExplorer.

6 Start ArcExplorer, if necessary.

ArcExplorer

7 Using Windows Explorer, drag and drop the downloaded file into the ArcExplorer window. When prompted, save the data in the same directory as the ZIP file (e.g., *C:\homedata*). Since it has a default.AEP file, all the themes are added to your map view automatically. The streets data and flood data are turned on for you.

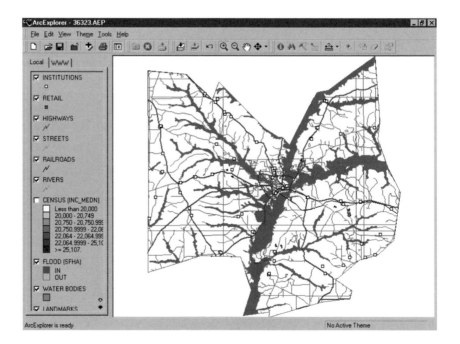

8 Now turn off the FLOOD theme. Turn on the CENSUS theme. It's symbolized according to income. Use this theme's other attributes to learn about
the people in your community.

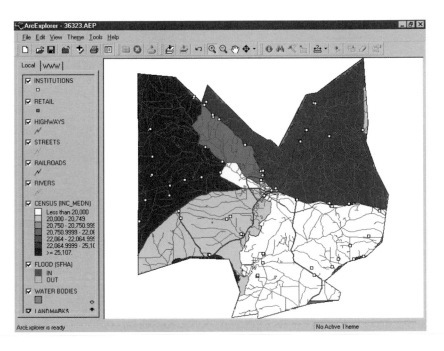

Be sure to check the *GIS for Everyone* Web site for other data sets in your
area of interest that are available for downloading.

Elsewhere on the Internet

The links provided on the *GIS for Everyone* Web site may be enough to get you started. But don't forget about the rest of the Internet, where thousands of gigabytes of geographic information are available. The amount of data may seem overwhelming, but it's quite easy to find what you need if you know where to look.

Search engines   Search engines are probably the best starting point for locating geographic data. Begin your search by accessing your favorite Web-based search engine. These include Excite (www.excite.com), AltaVista (www.altavista.com), Yahoo (www.yahoo.com), Lycos (www.lycos.com), HotBot (www.hotbot.com), and Infoseek (www.infoseek.com). Another popular search engine is MetaCrawler (www.metacrawler.com), which searches the other search engines, combining the results into one list, ranked by relevance. Different search engines will locate different sites, and each engine has its own unique features, so it pays to try them all to find those that suit you.

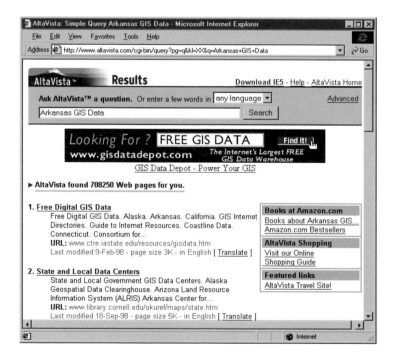

Try using combinations of keywords related to the specific data you're looking for. Include keywords to indicate that you're looking for GIS data, for example "GIS data," "spatial data," and "geographic data." Include keywords to indicate the geographic location. You could also include keywords to indicate the format you're looking for, such as "shapefile," "DRG," "DOQ," or "E00." Use different combinations of these keywords.

You'll wind up with many links to explore. Perhaps the most useful sites are those that specialize in providing collections of links to sites offering GIS data to download. You'll also find the Web pages of federal, state, and local government agencies, many of which offer free downloadable data. Academic and commercial data providers will also turn up.

Compressed files   Because GIS data sets can be quite large, many of the data files on the Internet are distributed in compressed format. You can use utilities such as WinZip or PKUNZIP to uncompress the data. You can also use ArcExplorer itself.

Add a .zip archive directly to an ArcExplorer project and ArcExplorer will add its supported data types to your legend. If a default.aep file is present in the archive, that project will open automatically.

If you unzip a .zip archive of shapefiles into a folder that already contains shapefiles, all shapefiles in that folder will be added to the ArcExplorer legend.

Metadata   Metadata describes the content, quality, condition, and other characteristics of data. Sometimes metadata sites will contain a link to download data. Other times, especially for large data sets, you'll find information about who to contact to obtain the data.

Interactive mapping sites

In addition to the wealth of GIS data available online, you'll also find many interactive mapping sites. These sites allow you to create maps without GIS software. You connect to the Web site using your browser and then compose a map while still online. When you're finished, you can either save the map as an image or send it to your printer. Some sites allow you to download the data you view for use in your own GIS.

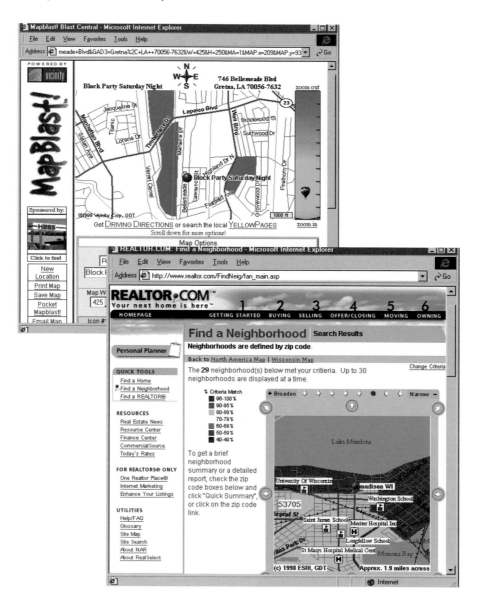

Exploration 13   Create a map of Sydney, Australia, using ArcExplorer as an Internet client

In previous explorations, all your data was accessed on your local computer. Now, if you've got an Internet connection, you'll use ArcExplorer as an Internet client. ArcExplorer allows you to view data directly over the Web on Internet sites set up for this. Some sites also allow you to download the data so you can use it locally.

In this exploration, you'll access world basemap data from a Web site and then download data for Sydney, Australia, to your local PC. (Again, you'll need to be connected to the Internet to complete this exploration.)

ArcExplorer™

1 Start ArcExplorer, if necessary.

2 Click the WWW tab at the top of the legend to switch to Web mode.

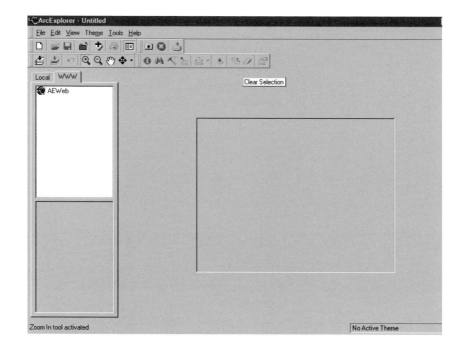

3 Click the Add Theme button.

The Open ArcExplorer Web Site dialog box displays. This is where you type in the URL of a Web site you're interested in. The first time you use ArcExplorer, the ESRI ArcExplorer Web Site is listed.

If you find and enter a URL other than ESRI ArcExplorer Web Site, ArcExplorer prompts you to save it as one of your AEWeb Favorites. You can choose Yes to save the URL and then enter a name, or choose No to open it without saving it as an AEWeb Favorite.

4 Choose ESRI ArcExplorer Web Site and click Add URL.

The ArcExplorer Web site contains U.S. and world geographic data for you to view and download. When you first open the site, you'll see an overview of either the United States or the world. As you zoom in, more detailed layers, such as country boundaries, city points, rivers, roads, railroads, and airports, are displayed.

Under AEWeb in the legend, you see the names of the data servers available on the ArcExplorer Web site.

5 Click *World BaseMap* to open the world data themes.

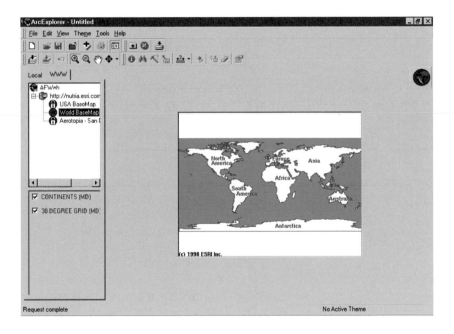

Since you want to display Sydney, Australia, use the Zoom In tool to change the map view.

6 Click the Zoom In tool.

Drag a box over southeastern Australia.

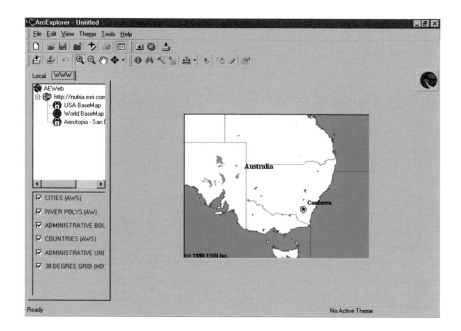

Notice that the map view redraws to show you the area in the box you defined. As you zoom in, layers of more detail draw. Once you've zoomed in to eastern Australia, you should be able to see Australia's capital, Canberra.

7 Continue to use the Zoom In tool until you have an extent that shows Sydney.

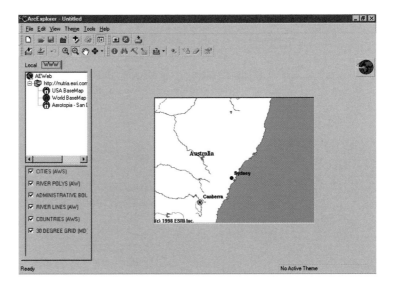

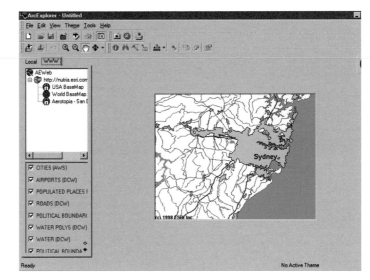

Once you have Sydney displayed, you'll retrieve the data for this area over the Internet to your PC.

8 Click the Retrieve Data from WWW tool. In the Retrieve Options dialog box, click Retrieve to download data for the current extent.

Retrieve Options

From the list below, select the amount of data to retrieve.

Select Your Extent

- ● The current extent
- ○ 2X the extent
- ○ 4X the extent

Retrieve

Cancel

At this point, ArcExplorer will try to connect to the map server—the machine on the Internet that's serving up the data. It may take a moment to load. If you get tired of waiting, you can stop by clicking the Cancel WWW Request tool. (Once ArcExplorer has located the map server, this button is disabled.)

9 Click Yes to accept the license agreement.

ArcExplorer

ESRI is willing to license the selected data to you only upon the condition that you accept all of the terms and conditions contained in the ESRI Data License Agreement found at http://www.esri.com/data/online/datalicense.html. By downloading the online data,

you are indicating your acceptance of the ESRI data license agreement. If you do not agree to the terms and conditions as stated, then ESRI is unwilling to license the data to you and you should not download the selected data.

Layer Description - (Shapefile Name)

30 Degree Grid (MD) - (mdworld3)
Political Boundaries Polys (DCW) - (ponetp)
Water (DCW) - (dnnetl)
Water Polys (DCW) - (dnnetp)
Political Boundaries Lines (DCW) - (ponetl)
Roads (DCW) - (rdline)
Cities (AWS) - (awscitie)
Populated Places Polys (DCW) - (pppolyp)
Airports (DCW) - (aepoint)

Click Yes to proceed with the download of these layers.

Yes   No

10 Choose or create a folder to which to download the data. (For example, *C:\Sydney* would do fine.) Click the Open button to download the data to your PC.

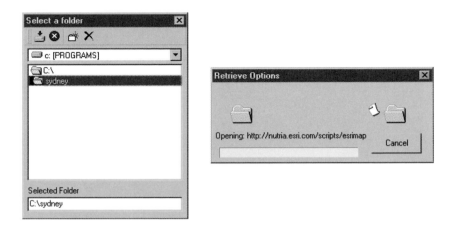

11 After the data has downloaded, click Yes to add the data to your local map view.

Now you can view the data that's been added to your local map view.

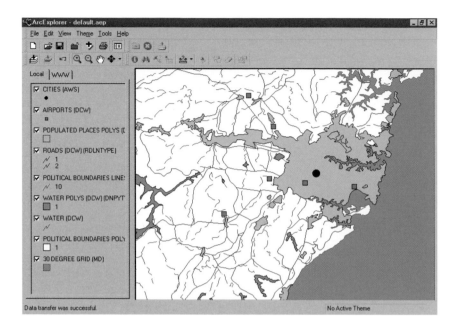

You've now downloaded the data into your map view and have a clipped extent of all the themes you were viewing in WWW mode. You also have a project named default.aep that you can open later.

12 From the File menu, choose Exit.

A list of Web sites enabled for ArcExplorer is available at the *GIS for Everyone* site.

Just the beginning

You now understand what digital data is and how to explore digital maps. You know how to ask questions and get answers from the databases that power those maps. You know how to create colorful and informative maps to print or to include in Web pages. And most importantly, you know how to find digital geographic data to create your maps.

After reading this book, you may be happy just knowing that GIS is making a difference in the world, and that it's available to you if you ever need it.

Or you might start making your own maps for school, work, community service, or just for fun. You might even introduce your family, friends, and neighbors to GIS by sharing your maps with them.

You may even start to think how GIS can play a role in your work and your community. ArcExplorer is just an entry point to this technology. More powerful software exists that can do some unbelievable things. If you want to do more, you might want to research GIS software such as ArcView® GIS and ArcInfo. The ESRI Web page is your starting point for this at **www.esri.com**.

You may even want to consider a career in GIS. Today, GIS is a multi-billion-dollar industry, full of opportunity and employing hundreds of thousands of people worldwide. GIS is taught in schools, colleges, and universities throughout the world. You can even get online training in GIS from the ESRI Virtual Campus at **campus.esri.com**.

Whatever you decide with your new-found knowledge, we wish you happy exploring.

ESRI license agreement

IMPORTANT

READ CAREFULLY BEFORE ELECTING TO DOWNLOAD AND/OR INSTALL THE SOFTWARE

ENVIRONMENTAL SYSTEMS RESEARCH INSTITUTE, INC. (ESRI), IS WILLING TO LICENSE THE SOFTWARE AND RELATED MATERIALS TO YOU ONLY ON THE CONDITION THAT YOU ACCEPT ALL OF THE TERMS AND CONDITIONS CONTAINED IN THIS ESRI LICENSE AGREEMENT. PLEASE READ THE TERMS AND CONDITIONS CARE-FULLY. THE SOFTWARE AND RELATED MATERIALS WILL NOT DOWNLOAD AND/OR INSTALL ONTO YOUR COMPUTER SYSTEM UNTIL YOU HAVE MANIFESTED YOUR ASSENT TO THE TERMS AND CONDITIONS OF THE ESRI LICENSE AGREEMENT. YOU ARE INDICATING YOUR ACCEPTANCE OF THE ESRI LICENSE AGREE-MENT. IF YOU DO NOT AGREE TO THE TERMS AND CONDITIONS AS STATED, THEN ESRI IS UNWILLING TO LICENSE THE SOFTWARE AND RELATED MATERIALS TO YOU. IN WHICH EVENT, THE SOFT-WARE AND RELATED MATERIALS WILL NOT DOWNLOAD AND/OR INSTALL ONTO YOUR COMPUTER SYSTEM.

ESRI license agreement

This is a license agreement and not an agreement for sale. This license agreement (Agreement) is between the end user (Licensee) and Environmental Systems Research Institute, Inc. (ESRI), and gives Licensee certain limited rights to use the proprietary ESRI® desktop software and software updates, sample data, online and/or hard-copy documentation and user guides, including updates thereto, trademark emblem, and trademark usage guidelines, as applicable (hereinafter referred to as "Software and Related Materials"). All rights not specifically granted in this Agreement are reserved to ESRI.

Reservation of Ownership and Grant of License: ESRI and its licensor(s) retain exclusive rights, title, and ownership of the copy of the Software and Related Materials licensed under this Agreement and, hereby, grant to Licensee a personal, nonexclusive, nontransferable license to use the Software and Related Materials based on the terms and conditions of this Agreement. Further, ESRI grants Licensee a personal, nonexclusive, limited trademark license to use the "GIS by ESRI" trademark emblem under the terms set for the below and in the *ESRI Trademark Usage Guidelines* for so long as this Agreement is in effect. From the date of receipt, Licensee agrees to use reasonable effort to protect the Software and Related Materials from unauthorized use, reproduction, distribution, or publication.

Copyright: The Software and Related Materials are owned by ESRI and its licensor(s) and are protected by United States copyright laws and applicable international laws, treaties, and/or conventions.

Third Party Beneficiary: ESRI is authorized to redistribute and (sub)license third party (hereinafter referred to as "Third Party Beneficiary") software and data component(s) that is (are) delivered with the Software and Related Materials provided under this Agreement. The Third Party Beneficiary shall have the right to enforce this Agreement to the extent permitted by applicable law.

Permitted Uses:

- Licensee may install the Software and Related Materials onto the permanent storage device(s) on the computer system(s) and use the Software and Related Materials for Licensee's own internal use.

- Licensee may make routine computer backups of the Software and Related Materials and one (1) archival copy of the Software and Related Materials during the term of this Agreement.

- Licensee may use, copy, alter, modify, merge, reproduce, and/or create derivative works of the online documentation for Licensee's own internal use. The portions of the online documentation merged with other Software, hard-copy, and/or digital materials shall continue to be subject to the terms and conditions of this Agreement and shall provide the following copyright attribution notice acknowledging the proprietary rights of ESRI and its licensor(s) in the online documentation: "Portions of this document include intellectual property of ESRI and its licensor(s) and are used herein by permission. Copyright © 199_ Environmental Systems Research Institute, Inc., and its licensor(s). All Rights Reserved." [Insert actual copyright date(s) from the source materials.]

- Licensee may reproduce and redistribute the Software and Related Materials provided all of the following occur: (1) the Software and Related Materials are reproduced and redistributed in their entirety whether as a stand-alone application or as an embedded GIS viewing tool on Licensee's own deliverable, (2) this ESRI License Agreement accompanies each copy of the Software and Related Materials and the recipient agrees to be bound by these terms and conditions, (3) all copyright and trademark attribution/notices are reproduced, (4) there is no charge or fee attributable to the use of the Software and Related Materials, and (5) the "GIS by ESRI" trademark emblem (globe.bmp or globe.jpg) that is included with this Software executable is included on the packaging and/or CD–ROM artwork for the Licensee's own deliverable, subject to the *ESRI Trademark Usage Guidelines* included with the Software executable.

Uses Not Permitted:

- Licensee shall not sell, rent, lease, sublicense, lend, assign, or time-share, in whole or in part, the Software and Related Materials for any commercial redistribution purposes. In the event that Licensee wants to commercially redistribute the Software and Related Materials, Licensee shall contact ESRI in order to negotiate an appropriate distribution agreement for such commercial use.

- Licensee shall not reverse engineer, decompile, or disassemble the Software or make any attempt to unlock or bypass the software keycode and/or hardware key used, as applicable, subject to local law.

- Licensee shall not remove or obscure any ESRI or ESRI licensor(s) copyright, trademark, and/or proprietary rights notices.

- Licensee shall not alter or modify the ESRI Software or prepare any derivative works from ESRI Software EXCEPT that Licensee may alter, modify, or create macros or scripts supported by the Software's macro or scripting language, as applicable.

Term: The license granted by this Agreement shall commence upon Licensee's receipt of the Software and Related Materials and shall continue until such time that (1) Licensee elects to discontinue use of the Software and Related Materials and terminates this Agreement or (2) ESRI terminates for Licensee's material breach of this Agreement. Upon termination of this Agreement in either instance, Licensee shall return to ESRI the Software and Related Materials, and any whole or partial copies, codes, modifications, and merged portions in any form. The parties hereby agree that all provisions, which operate to protect the rights of ESRI and its licensor(s), shall remain in force should breach occur.

Warranty Disclaimer: THE SOFTWARE AND RELATED MATERIALS CONTAINED THEREIN ARE PROVIDED "AS IS," WITHOUT WARRANTY OF ANY KIND, EITHER EXPRESS OR IMPLIED, INCLUDING, BUT NOT LIMITED TO, THE IMPLIED WARRANTIES OF MERCHANTABILITY AND FITNESS FOR A PARTICULAR PURPOSE. ESRI DOES NOT WARRANT THAT THE OPERATION OF SOFTWARE AND RELATED MATERIALS WILL BE UNINTERRUPTED OR ERROR FREE. IN THE EVENT ANY SAMPLE DATA HAVE BEEN PROVIDED HEREIN, THE SAMPLE DATA HAVE BEEN OBTAINED FROM SOURCES BELIEVED TO BE RELIABLE, BUT THEIR ACCURACY AND COMPLETENESS, AND THE OPINIONS BASED THEREON, ARE NOT GUARANTEED. EVERY EFFORT HAS BEEN MADE TO PROVIDE ACCURATE SAMPLE DATA IN THIS PACKAGE. THE LICENSEE

ACKNOWLEDGES THAT THE SAMPLE DATA MAY CONTAIN SOME NONCONFORMITIES, DEFECTS, ERRORS, AND/OR OMISSIONS. ESRI AND/OR ITS LICENSOR(S) DO NOT WARRANT THAT THE SAMPLE DATA WILL MEET LICENSEE'S NEEDS OR EXPECTATIONS, THAT THE USE OF THE SAMPLE DATA WILL BE UNINTERRUPTED, OR THAT ALL NONCONFORMITIES CAN OR WILL BE CORRECTED. ESRI AND/OR ITS LICENSOR(S) ARE NOT INVITING RELIANCE ON THESE SAMPLE DATA, AND LICENSEE SHOULD ALWAYS VERIFY ACTUAL SPATIAL AND/OR TABULAR DATA AND/OR INFORMATION. THE SAMPLE DATA CONTAINED IN THIS PACKAGE ARE SUBJECT TO CHANGE WITHOUT NOTICE.

Limitation of Liability: IN NO EVENT SHALL ESRI AND ITS LICENSOR(S) BE LIABLE FOR COSTS OF PROCUREMENT OF SUBSTITUTE GOODS OR SERVICES, LOST PROFITS, LOST SALES OR BUSINESS EXPENDITURES, INVESTMENTS, OR COMMITMENTS IN CONNECTION WITH ANY BUSINESS, LOSS OF ANY GOODWILL, OR FOR ANY FOR DIRECT, INDIRECT, SPECIAL, INCIDENTAL, AND/OR CONSEQUENTIAL DAMAGES RELATED TO LICENSEE'S USE OF THE SOFTWARE AND RELATED MATERIALS, HOWEVER CAUSED, ON ANY THEORY OF LIABILITY, EVEN IF ESRI IS ADVISED OF THE POSSIBILITY OF SUCH DAMAGE. THESE LIMITATIONS SHALL APPLY NOTWITHSTANDING ANY FAILURE OF ESSENTIAL PURPOSE OF ANY LIMITED REMEDY.

Waivers: No failure or delay by ESRI or its licensor(s) in enforcing any right or remedy under this Agreement shall be construed as a waiver of any future or other exercise of such right or remedy by ESRI or its licensor(s).

Order of Precedence: Any conflict and/or inconsistency between the terms of this Agreement and any FAR, DFAR, purchase order, or other terms shall be resolved in favor of the terms expressed in this Agreement, subject to the U.S. Government's minimum rights unless agreed otherwise.

Export Regulations: Licensee acknowledges that this Agreement and the performance thereof are subject to compliance with any and all applicable United States laws, regulations, or orders relating to the export of computer software or know-how relating thereto. ESRI Software and Related Materials have been determined to be Technical Data under United States export laws. Licensee agrees to comply with all laws, regulations, and orders of the United States in regard to any export of such Technical Data. Licensee agrees not to disclose or reexport any Technical Data received under this Agreement in or to any countries for which the United States Government requires an export license or other supporting documentation at the time of export or transfer, unless Licensee has obtained prior written authorization from ESRI and the U.S. Office of Export Control. The countries restricted at the time of this Agreement are Cuba, Iran, Iraq, Libya, North Korea, and Sudan.

U.S. Government Restricted/Limited Rights: Any Software and Related Materials delivered hereunder are subject to the terms of the ESRI License Agreement. In no event shall the U.S. Government acquire greater than RESTRICTED/LIMITED RIGHTS. At a minimum, use, duplication, or disclosure by the U.S. Government is subject to restrictions as set forth in FAR §52.227-14 Alternates I, II, and III (JUN 1987); FAR §52.227-19 (JUN 1987) and/or FAR §12.211/12.212 (Commercial Technical Data/Computer Software); and DFARS §252.227-7015 (NOV 1995) (Technical Data) and/or DFARS §227.7202 (Computer Software), as applicable. Contractor/Manufacturer is Environmental Systems Research Institute, Inc., 380 New York Street, Redlands, California 92373-8100 USA.

Governing Law: This Agreement is governed by the laws of the United States of America and the State of California without reference to conflict of laws principles.

Entire Agreement: The parties agree that this constitutes the sole and entire agreement of the parties as to the matter set forth herein and supersedes any previous agreements, understandings, and arrangements between the parties relating hereto and is effective, valid, and binding upon the parties.

ESRI is a trademark of Environmental Systems Research Institute, Inc., registered in the United States and certain other countries; registration is pending in the European Community. The ESRI globe logo and GIS by ESRI are trademarks of Environmental Systems Research Institute, Inc.

Field definitions for census data

WORKING WITH GIS means dealing with data files, themes, and data fields. It means navigating code-oriented waters. For example, you went looking for income data and came back with geographic data like CENSUS.SHP or field names like INC_MEDN and INC_3550. What are they? Do the data focus on countries or counties? For what year are the data? Where is the data from?

This appendix will help you locate and understand the census data you used in chapters 3 and 4. In the section that follows, you'll find listings that provide detailed descriptions and definitions for the fields associated with the census theme. The dictionaries furnish field names in forms recognizable as a computer database format (e.g., INC_MEDN) and translate them into English (e.g., *Median household income*).

Except where otherwise noted, all fields represent absolute counts of persons or housing units as they fall within particular categories. For example, persons can be counted according to ethnicity, and housing units can be grouped according to the number of rooms in the unit.

Population characteristics

Total population	Persons	Total population
	Families	Total families
	Houshold	Total households
	Samp_pop	Sample population
	Per_samp	Percentage of persons in sample
Sex	Male	Males
	Female	Females
Ethnicity	White	Whites
	Black	Blacks
	Amind	Native Americans
	Asian	Asian and Pacific
	O_ethnic	Other ethnicity
	Hispanic	Hispanics of all races
Age	Agelt_05	Age less than 5
	Age05_09	Age 5 to 9
	Age10_14	Age 10 to 14
	Age15_17	Age 15 to 17
	Age18_19	Age 18 to 19
	Age15_19	Age 15 to 19
	Ageis_20	Age 20 years
	Age21_24	Age 21 to 24
	Age20_24	Age 20 to 24
	Age25_34	Age 25 to 34
	Age35_44	Age 35 to 44
	Age45_54	Age 45 to 54
	Age55_64	Age 55 to 64
	Age65_74	Age 65 to 74
	Age75_84	Age 75 to 84
	Agegt_84	Age over 84

Age (continued)	Agege_03	Age 3 or older
	Agege_05	Age 5 or older
	Agege_16	Age 16 or older
	Agege_25	Age 25 or older
Marital status	Single	Never married
	Married	Married
	Separate	Married and separated
	Widowed	Widowed
	Divorced	Divorced
Family type	Oneperhh	One-person households
	Marrwich	Married-with-child families
	Marrnoch	Married-no-child families
	Singwich	Single-with-child families
	Singnoch	Single-no-child families
Persons in families	Perinfam	In families
Persons in institutions	Corrinst	In correctional facilities
	Nurshome	In nursing homes
	Mental	In mental institutions
	Juvenile	In juvenile detention
	O_instit	In other institutions
	O_noinst	In other noninstitutional group quarters
	Dormitor	In dormitories
	Miliquar	In military quarters
	Shelters	In homeless shelters
	Instreet	On the street
Ability to speak English	Onlyengl	Speak English only
	Ableengl	Able to speak English
	Cantengl	Cannot speak English

Citizenship	Native	Native citizens
	Natural	Naturalized citizens
	Noncitiz	Noncitizens
Region of birth	Born_ins	Born in same state
	Bornnort	Born in northeast USA
	Bornmidw	Born in midwest USA
	Bornsout	Born in south USA
	Bornwest	Born in west USA
	Born_for	Born outside USA
Migratory status	Samehous	At same house 1985
	Samecnty	In same county 1985
	Samestat	In same state 1985
	Oth_stat	In other state 1985
	Frm_abrd	In other country 1985
Workers	Workers	Total workers over 16
Place of work	Wrkincty	Worked in county of residence
	Wrkexcty	Worked outside county of residence
Type of commute	Drvalone	Drove alone to work
	Carpool	Carpooled to work
	Pubtrans	Public transit to work
	Wrk_home	Worked at home
Duration of commute	Com_lt15	Commute under 15 minutes
	Com_1529	Commute 15 to 29 minutes
	Com_3044	Commute 30 to 44 minutes
	Com_gt44	Commute over 44 minutes
Student population	Preprima	Preprimary students
	Elemscnd	Elementary and high school students
	College	College students

Educational attainment	Nodiplom	No high school diploma
	Highschl	High school graduates
	Somecolg	Attended college
	Colggrad	College graduates
	Gradschl	Attended graduate school
Employment status	Armdforc	In armed forces
	Employed	Total employed
	Unemploy	Total unemployed
	Notinwrk	Total not in workforce
Industry of employment	Primary	In primary industries
	Manufact	In manufacturing
	Utility	In transport and utilities
	Trade	In trade
	Fire	In finance and insurance
	Services	In commercial services
	Profserv	In professional services
	Public	In public administration
Occupation	Execprof	As managers and professionals
	Techsale	As technical/sales/clerical
	Service	As service employee
	Manual	As operator/assembler/laborer
Employment sector	Privwork	In private sector
	Publwork	In public sector
	Selfwork	Self employed
	Famiwork	Unpaid family work
Household income	Inc_lt15	Income under $15,000 in household
	Inc_1525	Income $15,000–25,000 in household
	Inc_2535	Income $25,000–35,000 in household
	Inc_3550	Income $35,000–50,000 in household
	Inc_5075	Income $50,000–75,000 in household

Household income (continued)	Inc_gt75	Income over $75,000 in household
	Inc_medn	Median household income
	Inc_aggr	Aggregate household income
Households by type of income	Wage_sal	Wages or salary in household
	Self_inc	Self-employed income in household
	Farm_inc	Farm income in household
	Inte_inc	Interest income in household
	Socs_inc	Social security in household
	Publ_inc	Public assistance in household
	Reti_inc	Retirement income in household
Per capita income	Incprcap	Per capita income
Poverty	Childpov	Children in poverty
	Inpovrty	Persons in poverty

Housing characteristics

Housing units	Housing	Housing units
Occupancy	Occupied	Occupied units
	Vacant	Vacant units
Tenure	Ownr_occ	Owner-occupied units
	Rent_occ	Renter-occupied units
Occasional units	Seasonal	Seasonal-use units
	F_migrnt	Migratory worker units
Rooms in unit	Room_1_3	Units with one to three rooms
	Room_4_6	Units with four to six rooms
	Room_gt6	Units with more than six rooms
Persons in unit	Prunit12	Units with one to two persons
	Prunit34	Units with three to four persons
	Prunitg4	Units with more than four persons
Housing value	Val_lt25	Units valued under $25,000
	Val_2550	Units valued $25,000–50,000
	Val_501c	Units valued $50,000–100,000
	Val_1c2c	Units valued $100,000–200,000
	Val_2c3c	Units valued $200,000–300,000
	Val_gt3c	Units valued over $300,000
	Val_medi	Median value of units
	Agg_valu	Aggregate value of units
Housing contract rent	Rnt_lt25	Unit rent under $250
	Rnt_2550	Unit rent $250–500
	Rnt_5075	Unit rent $500–750
	Rnt_751k	Unit rent $750–1,000
	Rnt_gt1k	Unit rent over $1,000
	Rnt_medi	Median rent

Units in structure	Detached	Single detached units
	Attached	Single attached units
	Duplex	Duplex units
	Apartmnt	Apartment units
Source of water	Pubwater	Water from public source
	Prvwater	Water from private source
Type of sewer	Pubsewer	Sewer with public utility
	Prvsewer	Sewer by private means
Age of housing units	Bltbfr70	Built before 1970
	Blt_7079	Built 1970–1979
	Blt_8084	Built 1980–1984
	Bltaft84	Built after 1984
	Medyrblt	Median year built
Source of heat	Publ_gas	Heat from gas utility
	Botl_gas	Heat from bottled gas
	Electric	Heat from electricity
	Fuelkero	Heat from oil or kerosene
	Coalwood	Heat from coal or wood
	Oth_heat	Heat from other source
Vehicles in household	Novehicl	No vehicles in household
	Vehicl_1	One vehicle in household
	Vehicl_2	Two vehicles in household
	Vehiclg2	More than two vehicles in household
	Aggvehcl	Aggregate vehicles

Data compatibility

CREATING DIGITAL MAPS often requires bringing together data from a wide variety of sources. It's important that you know as much as you can about the sources of your data, for several reasons. Two chief reasons involve map projection and map scale.

Themes must be of the same map projection

The earth's surface is curved, but maps are flat. To represent three-dimensional space on a two-dimensional surface, a mathematical transformation called a *projection* is used. Many different map projections exist to support a wide variety of uses; maps that you've seen hanging on the walls of classrooms are frequently in the Mercator projection. Projections are all distinguished by their ability to represent a particular portion of the earth's surface. A projection that gives an accurate depiction of one portion of the globe may not work for a different portion. For example, the Mercator projection is good for depicting the earth's surface at the equator, but at the cost of distorting features near the north and south poles. This is why Greenland looks bigger than it actually is on Mercator-projection maps.

If you're working with multiple data themes in ArcExplorer, they must all be in the same projection for you to be able to see them together. If they're not, they won't show up in the same view. If you have problems with this, the best thing to do is to find out from the source of the data whether the information is available in the map projection that suits your needs. You can often find out which type of map projection you're working with from the place you get the data. If you get it from a Web site, the projection will most likely be listed with its description, or in a Read Me file stored with the data.

Themes must be of the appropriate scale

In order to represent a portion of the earth's surface on a map, the area must obviously be reduced. This reduction is called map *scale*. It's defined as the ratio of map distance to ground distance, and is commonly expressed as a ratio or a fraction, such as 1:24,000.

The values on either side of the colon in this fraction represent the proportion between distance on the map and distance on the ground in the same units. For example, "1:24,000" means "1 map inch represents 24,000 ground inches," or "1 map meter represents 24,000 ground meters."

In general, small-scale maps depict large ground areas, but they show little detail. On the other hand, large-scale maps depict small ground areas, but show much greater detail. The features on small-scale maps more closely represent real-world features, because the extent of reduction is lower than that of large-scale maps. As map scale decreases, features must be smoothed and simplified, or not shown at all. In other words, a dime-sized lake on a large-scale map (1:1,200) would be less than the size of the period at the end of this sentence on a small-scale map (1:1,000,000).

Every data set is designed for display at a particular scale (or within a range of scales). For example, a 1:500,000 data set will look "right" when displayed at that scale, but will look too sketchy if displayed at 1:50,000. If displayed at 1:5,000,000, it will look too "busy" or crowded and will take too long to draw on the monitor. If you find themes from multiple sources, be sure they're compatible. Data gathered for display at 1:1,000,000 shouldn't be displayed with 1:10,000-scale data.

IMPORT71

A COMMON DATA FORMAT available on the Internet that you can use in ArcExplorer is the ArcInfo coverage. However, coverages are frequently in a portable format, E00, that you have to convert before you can display in your ArcExplorer project. The tool to do this, the IMPORT71 utility, is included on the CD.

INSTALL IMPORT71

1 Insert the *GIS for Everyone* CD in your CD–ROM drive.

2 Choose Run from the Start menu.

3 In the Command Line box, type the letter of your CD–ROM drive, a colon, a backslash, and import71.exe (for example, **e:\import71.exe**).

4 Either accept the default location to install IMPORT71, C:\Program Files\ESRI\Import71, or choose another one. Then click Next.

5 On the next screen, click Finish. When asked if you want to add a shortcut for this program to your desktop, choose Yes.

To use **IMPORT71**

Import71

1 Click on the IMPORT71 Utility icon on your desktop to bring up its dialog box.

> **Import71 Utility**
>
> Enter the name of the export file (include the 'e00' file extension). Then enter the name for the output data source.
>
> <u>E</u>xport Filename: [   ] <u>B</u>rowse...
>
> <u>O</u>utput Data Source: [   ] B<u>r</u>owse...
>
> <u>O</u>K   <u>C</u>ancel

The Export Filename is simply the full name and path of the .e00 file you wish to convert to a coverage, for example *H:\data\flood.e00*. You can type the location in but it's often easier to use the Browse button.

2 Click the Browse button to locate the .e00 file.

> **Open**
>
> Look <u>i</u>n: [austin]
>
> fema.e00
>
> File <u>n</u>ame: [fema.e00]   <u>O</u>pen
>
> Files of <u>t</u>ype: [Export Files (*.e00)]   Cancel

You need to use your PC's file management system to create a folder in which to store the coverage you'll create.

3 Using a program like Windows Explorer, create a new folder on your computer to store the coverage.

4 Type in the full path to your output data source. If you call the coverage *flood*, your path might be *C:\data\flood*.

You can also use the Browse button to choose the Windows Explorer directory to store the new imported coverage in, but you'll still need to enter a file name. Browse only lets you pick the directory to store it in.

5 When both values are filled in, click OK. The IMPORT71 Utility will import the .e00 file and create a coverage. It will inform you when it has finished.

Once the file has been imported, you can add it to your map view.

ARCEXPLORER TOOLBAR REFERENCE

Button	Name	Description
	New ArcExplorer	While ArcExplorer is running, starts an additional session of ArcExplorer.
	Open Project	Opens an ArcExplorer project (file with a .aep extension). (Local mode only.)
	Save Project	Saves an ArcExplorer project. (Local mode only.)
	Close Project	Removes all themes and returns an empty view. Closes the Web map site in WWW mode.
	Add Theme(s)	Adds one or more theme(s) to the view. Adds a Web map site in WWW mode.
	Print	Prints the map view and legend to a preformatted map layout. (Local mode only.)
	Toggle ArcExplorer Legend	Toggles the legend on and off.
	AEWeb Favorites	Opens the AEWeb Favorites dialog.
	Cancel WWW Request	Cancels a request to a map server for a download of WWW data.
	Retrieve Data from WWW	Downloads data displayed in the map view from WWW.
	Zoom to Full Extent	Zooms to the extent of all themes.
	Zoom to Active Theme	Zooms to the extent of the active theme. (Local mode only.)
	Zoom to Previous Extent	Zooms to the last previous extent. (Local mode only.)
	Zoom In	Zooms in on the position you click or the box you drag on the map view.

Button	Name	Description
	Zoom Out	Zooms out from the position you click or the box you drag on the map view.
	Pan	Pans the map as you drag the mouse across the map view.
	Direction	Chooses panning direction.
	Pan North	Pans the map view to the north.
	Pan South	Pans the map view to the south.
	Pan East	Pans the map view to the east.
	Pan West	Pans the map view to the west.
	Identify	Lists attributes of features you identify by clicking them in the map view.
	Find	Finds map feature(s) based on a text string you type in. (Local mode only.)
	Query Builder	Queries the active theme based on a query expression you construct.
	MapTips	Displays attribute information for features in the map view. (Local mode only.)
	Measure	Measures distances in the map view. You must first choose measurement units from the detachable menu.
	Address Match	Locates a street address or intersection in the map view.
	Clear Thematic Classification	Removes thematic classification from the active theme. (Local mode only.)
	Clear Selection	Clears the selected/highlighted features from the map view. (Local mode only.)
	Theme Properties	Sets the display characteristics of the active theme. (Local mode only.)

Books from ESRI Press

Enterprise GIS for Energy Companies
A volume of case studies showing how electric and gas utilities use geographic information systems to manage their facilities more cost effectively, find new market opportunities, and better serve their customers.   ISBN 1-879102-48-X

Transportation GIS
From monitoring rail systems and airplane noise levels, to making bus routes more efficient and improving roads, this book describes how geographic information systems have emerged as the tool of choice for transportation planners.   ISBN 1-879102-47-1

Getting to Know ArcView GIS
A colorful, nontechnical introduction to GIS technology and ArcView GIS software, this workbook comes with a working ArcView GIS demonstration copy. Follow the book's scenario-based exercises or work through them using the CD and learn how to do your own ArcView GIS project.   ISBN 1-879102-46-3

Serving Maps on the Internet
Take an insider's look at how today's forward-thinking organizations distribute map-based information via the Internet. Case studies cover a range of applications for Internet Map Server technology from ESRI. This book should interest anyone who wants to publish geospatial data on the World Wide Web. ISBN 1-879102-52-8

Managing Natural Resources with GIS
Find out how GIS technology helps people design solutions to such pressing challenges as wildfires, urban blight, air and water degradation, species endangerment, disaster mitigation, coastline erosion, and public education. The experiences of public and private organizations provide real-world examples. ISBN 1-879102-53-6

Zeroing In: Geographic Information Systems at Work in the Community
In twelve "tales from the digital map age," this book shows how people use GIS in their daily jobs. An accessible and engaging introduction to GIS for anyone who deals with geographic information. ISBN 1-879102-50-1

ArcView GIS Means Business
Written for business professionals, this book is a behind-the-scenes look at how some of America's most successful companies have used desktop GIS technology. The book is loaded with full-color illustrations and comes with a trial copy of ArcView GIS software and a GIS tutorial.   ISBN 1-879102-51-X

ARC Macro Language: Developing Menus and Macros with AML
ARC Macro Language (AML™) software gives you the power to tailor workstation ArcInfo software's geoprocessing operations to specific applications. This workbook teaches AML in the context of accomplishing practical workstation ArcInfo tasks, and presents both basic and advanced techniques. ISBN 1-879102-18-8

Understanding GIS: The ArcInfo Method (workstation ArcInfo)
A hands-on introduction to geographic information system technology. Designed primarily for beginners, this classic text guides readers through a complete GIS project in ten easy-to-follow lessons. ISBN 1-879102-00-5

ESRI Press publishes a growing list of GIS-related books. Ask for these books at your local bookstore or order by calling 1-800-447-9778. You can also shop online at www.esri.com/gisstore. Outside the United States, contact your local ESRI distributor.

ESRI Press ■ 380 New York Street ■ Redlands, California 92373-8100